Volume 3

ECONOMIC GEOGRAPHY

ECONOMIC GEOGRAPHY

B. W. HODDER AND ROGER LEE

Routledge
Taylor & Francis Group

LONDON AND NEW YORK

First published in 1974

This edition first published in 2015
by Routledge
2 Park Square, Milton Park, Abingdon, Oxon, OX14 4RN

and by Routledge
711 Third Avenue, New York, NY 10017

Routledge is an imprint of the Taylor & Francis Group, an informa business

British Library Cataloguing in Publication Data
A catalogue record for this book is available from the British Library

ISBN: 978-1-138-85764-3 (Set)
eISBN: 978-1-315-71580-3 (Set)
ISBN: 978-1-138-88498-4 (Volume 3)
eISBN: 978-1-315-71574-2 (Volume 3)
Pb ISBN: 978-1-138-88502-8 (Volume 3)

Publisher's Note
The publisher has gone to great lengths to ensure the quality of this reprint but points out that some imperfections in the original copies may be apparent.

Disclaimer
The publisher has made every effort to trace copyright holders and would welcome correspondence from those they have been unable to trace.

Economic Geography

B. W. HODDER and ROGER LEE

METHUEN & CO. LTD

First published 1974 by Methuen & Co. Ltd
11 New Fetter Lane, London EC4P 4EE
Reprinted 1977 and 1982

© *1974 B. W. Hodder and Roger Lee*

Printed in Great Britain by
Butler & Tanner Ltd, Frome and London

ISBN 0 416 07090 6

For my Parents   B.W.H.

For Lesley; Thomas,
who embodies all things good in my life;
and for my Mother   R. L.

Contents

The Field of Geography

Progress in modern geography has brought rapid changes in course work. At the same time the considerable increase in students at colleges and universities has brought a heavy and sometimes intolerable demand on library resources. The need for cheap text-books introducing techniques, concepts and principles in the many divisions of the subject is growing and is likely to continue to do so. Much post-school teaching is hierarchical, treating the subject at progressively more specialized levels. This series provides textbooks to serve the hierarchy and to provide therefore for a variety of needs. In consequence some of the books may appear to overlap, treating in part of similar principles or problems, but at different levels of generalization. However, it is not our intention to produce a series of exclusive works, the collection of which will provide the reader with a 'complete geography', but rather to serve the needs of today's geography students who mostly require some common general basis together with a selection of specialized studies.

Between the 'old' and the 'new' geographies there is no clear division. There is instead a wide spectrum of ideas and opinions concerning the development of teaching in geography. We hope to show something of that spectrum in the series, but necessarily its existence must create differences of treatment as between authors. There is no general series view or theme. Each book is the product of its author's opinions and must stand on its own merits.

W. B. MORGAN
J. C. PUGH
University of London
King's College
August, 1971

Figures

Tables

Acknowledgments

The authors and publisher wish to thank the following:

The Editor of *New Society* for figure 1.2
The University of Pennsylvania Press for figure 3.4
The Editor of *Regional Studies* for figure 3.6
George Bell & Sons Ltd for figures 3.7 and 3.9
Cambridge University Press for figure 3.10
John Wiley & Sons Inc for figures 5.1, 5.3 and 5.6
Weidenfeld & Nicolson Ltd for figures 5.7 and 7.1
The Editor of *Economic Geography* for figure 5.9
Pergamon Press Ltd for figure 6.5
George Allen and Unwin for figure 7.2
The Association of American Geographers for figures 1.3 and 7.3
Oxford University Press for figure 8.1
The Editor of the *Geographical Magazine* for figure 8.2
The Colston Research Society for figure 7.6
The Institute of British Geographers for figure 7.7
The University of Oregon Press for figure 8.3
Yale University Press for figure 9.1
Harper & Row Publishers, Inc for figures 3.3, 3.8 and 9.3
Prentice-Hall Inc for figures 9.4 and 9.5
The Editor of *The Canadian Geographer* for figure 9.5
Earth Island (Publishers) Ltd for figure 10.2

Preface

With the recent rapid growth of information in all branches of economic geography it might reasonably be argued that the attempt to deal with the subject as a distinct field of study is an impossible and perhaps intellectually futile task. A glance at any major series of university textbooks reveals how wide-ranging are the interpretations placed upon the subject matter of economic geography by its modern practitioners and demonstrates that generalization within the field is retreating rapidly with the advance of well-founded specialization. What then is the justification for a short introductory text on the whole field of economic geography?

The operation of economies is a vital and fundamental influence upon social structure and the distribution of power in society. This influence stems from the essentially integrated nature of economies, and is shaped by the social value system which is both a cause and consequence of prevailing economic values. In our own teaching of economic geography we have found it fruitful to build around the neo-classical concept of the economy as a framework upon which to construct the otherwise diffuse and discrete subject matter. Such a concept also points to the dominance of neo-classical economic theory (with its implied value system) in the theoretical development of economic geography. Furthermore, the concept enables us to begin to discern the importance of economic power as the key to the complex relationships between economy and society. Whilst it is not possible in a review of contemporary economic geography, based as it is upon neo-classical theory, to develop this theme very fully, it is hoped that the explicit recognition of the bias and inherent limitations of the neo-classical approach will clear the ground for a more realistic economic geography in the future.

Thus while the following pages introduce the reader to many lines of thought in the literature of economic geography, the major aim of this book is to tie these ideas together within the concept of the economy so as to provide a simple and logical basis for discussion, further reading and subsequent specialization. Many will disagree with some of the ideas, interpretations or emphases given here. But if a short introductory book of this kind is to have any value at all it must express a clearly defined if rather personal point of view.

Many people have contributed to this book, but our greatest debt must be to our students and teachers. Professor W. B. Morgan gave us much helpful criticism and prodded us when this was most needed. Dr Ronald Ng kindly read the proofs and made a number of valuable suggestions. Maureen List, Steve Pratt and Don Shewan drew the diagrams under Don's most calm and efficient supervision. Pete Newman and Alan Gillard converted the originals into prints. Linda Agombar, Mary Putney, Valerie Armes and Lesley Lee were compelled to combine their typing efforts to cope with our illegible manuscripts. We apologize once more, but remain most grateful for their patience.

London, 1974   B.W.H. and R.L.

Part 1
The economy and economic geography

In this section we present a simple model of the economy around which the rest of the book is constructed. The model also serves to indicate certain themes and issues arising from the geographical study of economies and these are considered against the background of a selection of the diverse literature in the field of economic geography.

1 A concept of the economy

In this book the term 'economy' refers to a network of economic decision makers. Such a mechanistic interpretation derives from neo-classical economics as the major source of theory in economic geography. This dependence is misplaced as neo-classical economics hides the fundamental structural relationships between *people* in the social production of material life beneath its concern with the exchange of commodities considered merely as material *objects*. However just as economics is breaking free from its neo-classical strait-jacket so too should economic geography. The speed with which it does this is an important determinant of the rate of obsolescence of this book.

Underlying the argument pursued in these pages then, is the neo-classical concept of the economy which, so it seems to us, can only be fully understood when treated as a whole. But economies, as defined here, can exist at any scale. They range from simple subsistence village economies through to those operating at the national or international scale. Yet even at its simplest the economy is a complex phenomenon. The reader may find it useful to consult some of the descriptive case studies of simple village economies written by economists, social anthropologists and geographers (e.g. Firth 1952; Epstein 1962; and Hill 1972). It is not possible to translate their findings from one spatial scale to another, but such studies are useful in that they introduce the concept of the economy by describing, within the context of a small and specific community area, the complex integrated and continually changing interaction of variables affecting economic activity. In this book, however, the approach is to try to simplify the characteristics and operation of economies, at whatever scale, down to common elements and processes. We begin by setting up a very simple model of the economy.

A simple model of the economy
The economy, as revealed in fig. 1.1, may be said to consist of:

1. A set of decision-making elements identified as consumers (E_1); firms (E_2), often grouped into particular industries or sectors; resource owners (E_3); and government (E_4).
2. A set of relationships between the elements, shown in fig. 1.1 by the dashed and continuous arrows connecting them, facilitated by the

market for goods and services (M₁) and the market for factors of production (M₂).

3. A set of relationships between these components and the total environment, which may be simply defined as a higher order system within which the economy, or economic system, is but a part.

In this sense economic activity can be said to consist of the actions and interactions of four sets of elements. For a variety of reasons, the most basic of which is self-preservation, *consumers* (usually measured as consumer-households) generate wants and needs or demands for goods

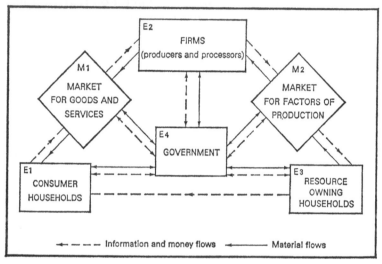

1.1 *The simple economy.*

and services. *Firms*, or the technical and organizational units for production and processing, may respond to this demand by deciding to produce the goods demanded. To do this they need to acquire the resources of land, labour and capital from *resource owners* (resource-owning households). In other words, firms may generate a demand for resources and may try to attract them by the offer of some form of payment to the owners of the resources. If this offer is satisfactory the resource owners may decide to release at least some of their resources. Firms can then generate productive services (or factors of production) from the input of resources which are used in combination to make goods by the process known as production. The goods, or output produced by firms, can then be offered to the consumers. If the latter are satisfied with the goods and decide to accept them, then the firms receive a return, or payment, in exchange for the utility of the goods. In reality, all economically active households in the economy are both consumers and resource owners; the payment, or income, received from firms by resource owners may in turn be used to purchase goods from firms by the same individuals in their role as consumers. Transfer payments are made via *government* to those households unable, for reasons of old age, ill health or unemployment, to be economically active; such payments are supposed to enable these households to behave as consumers.

Firms are linked to consumers and resource owners along a network of communications via the *market for goods and services* and *the market for factors of production*. A market is a process of communication that enables buyers and sellers to exchange information about actual or latent demands and available or potential supplies, and helps them to organize the sale and purchase of goods and services. This type of activity may take place within well-defined market places which act as commercial foci for buyers and sellers and so are referred to as central places. But not all commodities and information are exchanged at a central place. The market for iron ore, for example, has no places of exchange at which buyers are able to meet and draw upon supplies from a particular market-supply area. Furthermore, the reduction in ore transportation costs is causing formerly discrete supply areas to coalesce, so that any one supplier of iron ore may well serve buyers in spatially separate locations (Manners 1971a). Sometimes the term market is used in a less specific context to include all the intermediate and final destinations of goods or resources being exchanged between buyers and sellers. In chapter 8 we shall devote more attention to the study of market *places* but our present interest in markets is as the means by which information and goods are exchanged within economies.

This description of economic activity assumes that it is set in motion and controlled by decentralized decision makers whose activities are stimulated by the likelihood of private gain and coordinated through the markets. Although there are good theoretical reasons for such an arrangement the theory is not easily transferred to the real world. Furthermore, when the essentially economic issues of social justice and ecological balance are considered, an economy based upon private decision making is sadly lacking (see pp. 6–10). Alternatively, economic activity can be generated and controlled by a single central decision-making body – the government. All decisions about demands, supplies, the use of resources and the behaviour of the individual elements may be contained in a centrally controlled economic plan and the government, whose interests in economic activity are less private than are those of the individual elements of an economy, may take decisions on an entirely different basis. More common than either of these two extremes is a mixture of decentralized and centralized (private and public) decision making in which the government can influence individual actions; act as a consumer, producer or resource owner; and can outline and implement a coordinated economic plan.

Measuring the flows around the economy

Just as the individual household or firm may record its receipts and expenditures, so too does the economy record its receipt of income and its expenditure upon goods and services. Because economies are most effectively cordoned along a national political boundary, where international economic transactions can be recorded with comparative ease, the most highly developed accounting systems are at a national scale. Regional accounts are sometimes constructed, but regions are far more open to outside influences than are nations and, in any case, do not usually have any effective measuring cordon around them.

One of the most fundamental measures of the size of a national economy is its gross national product (G.N.P.) defined as the total market value of finished goods and services produced by the nation over a year. Defined in this way G.N.P. is the sum of all spending by consumers, investors and government together with net exports – the difference between what is sold and bought abroad. G.N.P., however, can also be considered from the income or cost side which, because of the circular flow of money around the economy, must equal G.N.P. defined in terms of expenditure. In this second sense G.N.P. is the total cost of production and thus represents the sum of all wages, interest, profits, depreciation and sales taxes. G.N.P. minus all income from abroad is known as gross domestic product (G.D.P.). For comparative purposes it is sometimes appropriate to subtract depreciation from G.N.P., leaving net national product (N.N.P.), which is a measure of output excluding that for capital replacement. By subtracting sales tax from N.N.P. we are left with the value of the production of final, usable goods and services, known collectively as national income. By contrast the personal income received by individuals, involves certain additions and subtractions from national income. Thus profits retained by firms, taxes on company profits and compulsory social security payments must be subtracted, whilst interest on government loans and transfer payments must be added to national income in order to arrive at personal income. But, as is all too well-known, personal income does not become disposable income until income taxes have been paid.

The environmental relations of the economy

As pointed out on p. 4, the economy cannot exist in isolation, being but part of the total wider environment. We consider here some of the external social and ecological relations of the economy.

SOCIAL RELATIONS

It can legitimately be argued that a study of the economy as a mere generator of economic activity is both socially irrelevant and conceptually unsound since one of the major problems facing societies, at all levels, is the social implications of the concentration of economic power. This may be defined as the power of economies or their individual elements to influence, by unilateral action, the course of economic events in a lasting or substantive fashion. The international inequalities in economic development are very well-known, one implication being that less than 25% of the world's population generates over 60% of its output (table 1.1). The remaining 75% of the population is, in consequence, placed at a considerable bargaining disadvantage in attempting to obtain a share of the world economy's productive capacity and output commensurate with its needs. But some would go further than this and argue that the very process of development in one part of the world has the effect of impoverishing other parts because developed areas operate at higher levels of economic efficiency and so undercut less efficient producers. Furthermore, by offering higher remuneration, developed economies suck in resources from less-developed economies and so increase their own productive potential while decreasing the potential of the source areas. Without an efficient mechanism for redistributing such spatially concentrated wealth, major and increasing inequalities in social well-being develop. On the other hand, in the implementation (or perhaps more correctly imposition) of development programmes it is important not to transfer value systems from one situation to another. It is all too easy, for instance, to assume that the dream of each less-developed country is to become developed. Such a view can easily disregard highly developed local cultures. Self-respect is at least as important a measure of social and economic progress as are increases in urbanization, specialization, long-distance trading connections and material wealth (Adams 1970).

A non-spatial and more frequently publicised example of inequality in economic power is symbolized by the struggle for shares in the wealth created by capitalist production between, on the one hand, the owners and controllers of capital and firms and, on the other, the owners of labour resources. The inequalities in the economic power of these elements are related to the private ownership of capital and the ability of capitalist institutions to make resource-allocation decisions. Thus the distribution of income amongst resource owners, which is determined by the type, amount and firms' evaluation of their resources and by the organized strength of their bargaining position, is an obvious symptom of their economic power and determinant of their social status and political influence. Clearly, as later chapters will show, the work situation is a fundamental social variable of much wider significance than in the mere production of goods.

TABLE 1.1 World[1] population and economic output 1960–70

Economy[2]	Population				Gross domestic product					
	Size (10⁶)		% of world total		Size ($ × 10⁸)		% of market economies' total			% of world total
	1960	1970	1960	1970	1960	1970	1960	1963	1970	1963
Planned economies	331·5	369·7	14·3	12·9	comparable data not available					22·7
Market economies	1979·2	2 505·5	85·6	87·1	11 278·0	24 698·0	100·0	100·0	100·0	77·3
Developing	1345·0	1795·0	58·2	62·4	1841·0	3 866·0	16·3	16·0	15·6	
Developed	634·2	710·5	27·4	24·7	9437·0	20 832·0	83·7	84·0	84·3	
World	2 310·7	2 875·2	100·0	100·0						100·0

Sources: Derived from United Nations (1972 and 1973)

Notes: 1. Excluding China, Democratic People's Republic of Korea, Mongolia, Democratic Republic of Vietnam
2. As defined by United Nations (1972)

ECOLOGICAL RELATIONS

It has been frequently brought to public notice in recent years that economic activities are taking an increasing toll of balanced interactions within the life-giving ecosystem. The links between the economy and its ecologic environment are shown schematically in fig. 1.2. A

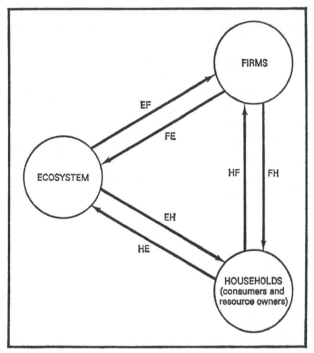

1.2 *The economy and the ecosystem. (Source: Coddington 1970.)*

circular, physical flow of materials can be traced out of the ecosystem through the economy and back into the ecosystem. This flow suggests that a crude policy of economic growth attacks the ecosystem in two ways: (i) by increasing flows of EF it depletes the natural resource base and feeds the flows of FE and HE which (ii) pollute the ecosystem with their waste and often toxic substances. It is now widely felt that the ability of the ecosystem to withstand this two-pronged attack is limited and that the effects of damaging withdrawals and additions are often intensified as they make themselves felt throughout the system as a whole.

The demands made upon the ecosystem by the economy are increasing for two main reasons. First, the continued and increasing expansion of world population creates new demands upon the productive capabilities of economies and is partly responsible for the ever more widespread use of ecologically dangerous technology, especially chemical technology, in the short-term interest of increasing the output of goods. Secondly, the increasing standard of living of the members of advanced industrial economies results in an expanding range of sophisticated consumer tastes, especially for personal mobility which is threatening remote, ecologically fragile stretches of wilderness, congesting and polluting urban areas, and consuming vast quantities of non-renewable petroleum reserves. Increases in the flows within an economy automatically lead to increased demands upon the ecosystem, the cost of which can only be measured by the extension of conventional accounting systems adept at recording flows of HF and FH to include EF, FE, EH and HE (fig. 1.2).

The interaction of the economy with its higher order society and ecosystem illustrates the general problem posed by the boundaries and environment of a system under analysis: which elements are within the boundaries of the system, and which are in the environment and judged not to have any relevance for the analysis? While our emphasis in these pages is very much upon the economy *per se*, it is important to bear in mind the crucial links which exist between the economy, society and the ecosystem.

The spatial structure of the economy

A similar problem of delimitation faces the spatial analyst of the economy. Where does it end and where begin? Firms, for example, may sell goods to local, regional, national or international groups of consumers and may acquire resources, especially of capital equipment, from an equally wide distribution of resource owners. How can the economic geographer delimit the economy to which the firm belongs? The simple answer is that he cannot. The world economy may be considered closed, but any attempt to delimit a sub-world economy in either spatial or sectoral terms must accept that the economy, so defined, is more or less open to interaction with other spatial units or sectors. This point is taken up again later on when dealing with economic interaction in chapter 7.

The spatial structure of an economy consists of the two-dimensional spread of the aspatial, functional economy discussed earlier and represented in fig. 1.1. Such a spatial arrangement may be considered at several levels or scales of analysis. First, the individual elements within a particular sector of the economy may be regarded as nodes of activity dependent upon, and interacting with, the other elements along a network of communications. Much theorizing about the location of firms has taken place at this scale, the problem being to minimize the cost of production and resource acquisition whilst at the same time trying to locate in such a way relative to consumers that profits are maximized.

Secondly, the individual elements may be coalesced into the space economies of cities and, at a larger scale, of world core areas forming zones of intense economic activity set in less intensively developed hinterlands. Although self-sufficient (or closed) in many respects, these concentrations need to draw upon the less densely distributed economic

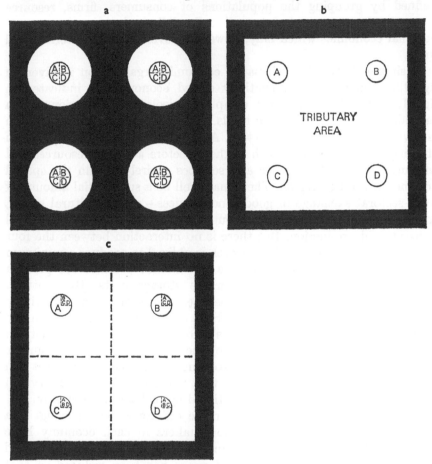

1.3 *Economic integration and locational specialization.*
(*Source: based on Webb 1959.*)

elements in their hinterlands for food and space-extensive resources and they interact with each other in the exchange of specialist products and information. The attempt to build a theory of the size, spacing and interaction of such economies, involving as it does the inter-dependence of a multitude of economic and social processes, is one of the most difficult problems in spatial economic analysis.

Both these levels of analysis are essentially functional in concept because they rely for their spatial definition upon the degree of closure of the economies. But closure is not necessarily determined by the existence of national political or regional-administrative boundaries. However, nation states and some supra-national groups of states are important units in economic activity, if only because most government intervention takes place at this scale. Thus, a third level of analysis is defined by grouping the populations of consumers, firms, resource owners and governments into politically delimited national or supra-national economies which may, however, develop strong international links.

A simple illustration of spatial economic organization is given in fig. 1.3. The concept of spatially closed economies is indicated in fig. 1.3a, which shows four completely isolated city-regions, each providing entirely for its own needs. The economy of each city-region is divided into four major sectors, A, B, C, D, located mainly within the major urban centre. Each of these sectors acquires resources and produces and markets goods and services entirely within the spatial confines of the city-region. There may well be a substantial amount of inter-sectoral exchange of goods and services – an agricultural sector, for example, may use some of the outputs of an engineering or chemical sector to aid production. But there is no interaction between the four city-regions, with the result that each is obliged to produce a minimum quantity of goods from all sectors but cannot produce too great a surplus without incurring considerable storage costs. By contrast, fig. 1.3b shows four completely open economies operating effectively as one highly integrated economy. Here each centre specializes on one economic sector which can draw its resources from within the city in which it is located, from all parts of the tributary area and from the other three cities. Furthermore, exchange of goods and services can take place between all the cities so that the demands in each city can be fully met despite complete locational specialization of supply.

Three major changes account for the differences between fig. 1.3a and b. First, the economic decision makers in each economy have perceived that there is advantage in local economic specialization and exchange; secondly, the friction of space has been reduced so that economic interaction over long distances is possible; and thirdly, organizations enabling the large-scale exchange of resources and goods and services over long distances have grown up. Fig. 1.3c shows an intermediate and more realistic situation. Specialization has occurred but is not complete, and whilst each economy can draw upon the whole of the tributary rural area, it is more probable that intra-sectoral inter-action will take place between elements close to rather than further from the town. But a precautionary note should be added here. Non-spatial factors may be more important than spatial factors in determining the nature of a space economy; for example, population growth may induce closure and economic diversity in the four economies by increasing economies of scale and by encouraging the production of an ever-increasing range of goods. Finally, before leaving fig. 1.3, it should be noted that although the discussion has been conducted here in terms of city-regions and sectors, the logic of the argument is not altered if nations or world core areas are substituted for the city-regions.

Conclusions

Three points can be made about this introductory and highly simplified description of the economy as a basis for the study of economic geography. First, it illustrates the essentially integrated nature of economies. Economic activity is generated by the individual elements of the economy – consumers, firms, resource owners and governments – and consists of their actions (e.g. the processes of demand-generation and consumption) and interactions (e.g. the exchange of information, commodities and payment). None of the individual elements is taken to have any meaning outside the context of the economy to which it belongs. In this sense the economy provides a focus and framework for the otherwise apparently discrete components of the subject matter of economic geography. At a very high level of generalization, integration of the kind noted here is common to all economies. Even a subsistence farmer within an almost closed and isolated economy is, in this sense, a much more complex economic man than is his equivalent within an open, highly specialized economy working, say, on a motor-vehicle assembly line. The subsistence farmer is simultaneously involved with the duties of a consumer, a firm, a resource owner and a government; moreover, he usually has not entirely separated out his 'economic' from his 'non-economic' activities. By contrast the 'economic activities' of the assembly worker are clearly distinguished and limited to consumption and resource marketing. Specialization of functions in a more developed economy has reduced the number of essential tasks that must be performed by its individual members, whilst the economy as a whole has become more complex as it must harmonize millions of individual decisions. By contrast, the task of harmonization in less specialized economies is accomplished, in large measure, by individuals and not by the economy as a whole.

Secondly, the notion of the economy focusses attention upon some of the wider issues and as yet unanswered questions with which economic geography is becoming increasingly concerned. Take, for instance, the relationships between spatial structure, economic efficiency and social justice. Do the spatial distributions of, say, shops or industry within a city discriminate economically between certain localities and groups of people? What are the transport cost implications of a policy of population redistribution? Do alternative spatial structures of economies affect economic efficiency? What are the social implications of spatially and personally concentrated ownership of resources? How can effective measures to conserve the ecosystem, involving the ecological modification of economic criteria for decision making and an understanding of the ecological role of the decision-making elements of an economy, be put into practice? Farmers, for example, cannot be considered merely as producers of agricultural commodities: they are also one of the most important groups of managers of the rural environment and their economic decisions can affect the ecological balance and landscape of vast areas.

Finally, the concept of the economy underlies the logic of the discussion in the sections that follow. The next chapter presents a selective review of literature in the field of economic geography against the background of the concept of the economy developed so far. Then follows the second and longer section of the book in which there is a discussion of the mechanisms of decision making and control. This involves a study of demand, supply, price and the market mechanisms; the alternative or complementary role of government control; and the decision-making activities of consumers, firms and resource owners. The third section of the book takes a closer look at the physical expression of integration within and between economies in the form of flows of goods, people and ideas, and the spatial structure of market centres and transport networks. The final section is devoted to a brief review of the geography of economic growth and development. This is a rapidly growing field of applied economic geography and demonstrates the apparent inability or unwillingness of the world economy to redistribute economic power or wealth from the rich to the poor. Intolerable economic inequalities, whether structural or spatial, and the associated ecological implications of the great but highly concentrated productive power of the world economy are perhaps the two major problems facing mankind.

2 Geographical studies of economic activity

The concept of the economy has rarely been used explicitly in geographical studies of economic activity; economic geography, in general, has been concerned with matters other than the operation of economies, the behaviour and interaction of their elements or the implications of the prevailing distribution of economic power. Within the existing literature in economic geography it is possible to distinguish two overlapping approaches to the study of economic activity. *Systematic* approaches are normally defined in terms of specific products (e.g. wheat), sectors (e.g. energy) or processes (e.g. trade), and are concerned with the spatial structure of these phenomena. *Spatial* approaches are defined in terms of specific two-dimensional, abstract, national or regional space and are concerned with the spatial structure of economic activity within these areas and with the effects of economic activity upon their economic and regional character.

Systematic approaches
An early emphasis in economic geography was upon the scientific study of world areas in their direct influence upon the production of goods (Götz 1882). This approach sought to lay bare the influence of the natural environment upon the occupations, products and lives of people in different regions of the world (Wooldridge and East 1966). The best-known example of this approach is *Chisholm's handbook of commercial geography* (1889) which remains an influential text now in its nineteenth (1975) edition. Chisholm considered both individual commodities and nations, and laid some stress upon generalized and descriptive discussions of the 'geographical' factors – by which was meant largely physical factors – affecting the production, transport and exchange of commodities. Newbigin's (1928) work fell into the same category, and Brown (1930), expressing a similar point of view, stated that economic geography was that aspect of the subject matter of geography which dealt with the 'influence of the environment, inorganic and organic, on the economic activities of man' (209). Much more recently, the same kind of viewpoint has been implied by the statement that 'economic geography, a hybrid science, is . . .

the systematic study of wealth which has its origins in the earth, oceans, seas and atmosphere' (Hope 1969, 1). On a more analytical level Paterson (1972) suggests that economic geography is 'concerned with the usefulness of earth features to man, with the amount of support they can afford him, and with the measures he may take to bring them into use' (1).

Perhaps the most influential single event which steered the course of systematic economic geography away from its antecedents in commercial geography was the publication of R. O. Buchanan's *The pastoral industries of New Zealand* (1935). In this work Buchanan explained his oft-quoted approach to economic geography: 'that, in industries organized on a commercial, as distinct from a purely subsistence basis, geographical conditions express themselves, if at all, in economic, mainly monetary terms; and that the geographical conditions are themselves dependent on the precise nature of the economic conditions. That will be recognized as merely specifying one type of instance of the generally accepted view that geographical values depend upon the cultural stage achieved by the human actors' (XV). In short, geographical values are not absolute, but are relative to the cultural environment – including the economic environment – in which they are located. From this basis, which also has its origins in the earlier work of Baker (1921 and 1925), has sprung a whole series of well-founded studies in systematic economic geography, including a large number of comprehensive texts (Carter and Dodge 1939; Thoman, Conklin and Yeates, 1968; Alexander 1963). Such texts present formidable obstacles for the reader because of the sheer weight of descriptive material contained within them. McCarty and Lindberg (1966) have tried to overcome this by providing a 'preface' to economic geography. By contrast the case-study approach (e.g. Rutherford, Logan and Missen 1966) provides general coverage with restricted specific examples. In addition, following the early example of Zimmerman (1933; 1972) there has appeared a number of texts on specific systematic topics strongly emphasizing the role of economic values in the structure and functioning of geographical patterns (Hall 1962; Martin 1966; Estall and Buchanan 1973; Manners 1971b; Odell 1963; Simpson 1966; Thoman and Conklin 1967; Berry 1967; Alexandersson 1967; Gregor 1970; Guyol 1971; Coppock 1971; Taaffe and Gauthier 1973; Hay 1973). Manners' (1971a) analysis of the world market for iron ore adopts the conceptual framework of economic analysis, and in so doing creates an integrated and comprehensive study based upon the analysis of the relationships between a natural resource and other elements of the world economy.

It is clear that in much of this systematic economic geography a conscious and most significant attempt is being made to bridge what

Wise (1956) has called the 'no-man's land' between geography and economics. The appearance of M. Chisholm's *Geography and economics* (1966, 2nd edn., 1970) was a landmark in the process of convergence between the two disciplines; this book expounds at a generalized scale of analysis 'the spatial implications of some parts of economics' (27). This is a more complex and difficult task than an examination of economic forces *per se*, so that Chisholm's book is perhaps most useful when used in conjunction with an introductory text in economics (Fleming 1969; Lipsey 1975; Samuelson 1977). Used in this way it provides an empirically based introduction to some major principles of economics of particular relevance to the spatial organization of capitalist economies.

It is also clear that this viewpoint of economic geography is quite explicitly spatially oriented, which is why there is a substantial overlap between the 'systematic' and 'spatial' approaches. Economic geography is interpreted by many of its practitioners (e.g. Smith, Taaffe and King 1968) as a study of the spatial pattern and location of economic activity. Central-place theory (Berry and Pred 1965) demonstrates how geographical theory may be derived from the basic postulates of economics: in this case elementary demand analysis is used to define a fundamental spatial concept – the range of a good. In a similar way, the economic idea of the margin has been modified and extended to form a basis for industrial location theory (Rawstron 1958; Smith 1971); whilst Chisholm (1968) has utilized von Thünen's work (Hall 1966) as a basis for the analysis of rural land use and settlement. Following Harvey (1966), who presented a wide-ranging review of social and economic theories bearing upon the spatial structure of agricultural land-use, Found (1971) has integrated these concepts from economics and behavioural science in a theoretical approach to rural land-use patterns. Similarly, Morgan and Munton (1971) have reviewed both economic and behavioural concepts in the study of agricultural geography.

Spatial approaches
The close interdependence between economic and political activity, and the fact that the governments of nation states are influential sources of power in relation to economic activity has frequently been noted (Fisher 1948). The state is able to control and modify the economic environment both within and to a certain extent beyond its boundaries in the pursuance of national or international policies (Chisholm 1970). Furthermore, the nation is important in economic terms as a unit of accounting and in social terms as a fundamental influence upon all aspects of life. It is hardly surprising therefore that many of the major pieces of work on the economic geography of spatial units have been undertaken at the scale of the nation state. Early important

B

work of this kind, employing economic analysis to full effect, is represented by Smith (1949) on the economic geography of Great Britain. This has been followed by a series of texts which include studies of parts of countries (Estall 1966), individual countries (Hamilton 1968; Courtenay 1972) and groups of countries (Mead 1958; O'Connor 1968; White and Gleave 1971). Other important recent texts employing a similar approach include an analysis of modern France (Thompson 1970), the E.E.C. (Parker 1968), Latin America (Cole 1965) and the United States (Estall 1977). The unique importance of the state and its planning machinery for economic activity in the U.S.S.R. has been demonstrated by Cole and German (1972) and in Yugoslavia by Hamilton (1968), whilst Tregear (1970) attempts to draw out the relationships between Chinese communism and the economic geography of China. At a more local scale, but still strongly emphasizing the role of government is a series of texts on the industrial geography of Great Britain (Smith 1969; House 1969; Lewis and Jones 1971; Humphrys 1972).

The convergence of regional and economic geography (Haggett 1970) has also been prompted by a wide-ranging interest in regional studies within the social sciences which has given rise to the rapid growth of regional science. This multi-disciplinary field of study has been very productive of theoretical concepts and techniques for spatial analysis (Garrison 1959, 1960). Economics has made the running in its development, being concerned with 'the economic implications of the spatial dimension' (Richardson 1969, 5) – a statement which contrasts interestingly with that of the geographer (quoted on p. 17) who refers to the spatial implications of economic activity. But economic geography has benefited from exposure to the wide range of relevant literature and rigorous methods of analysis (Isard et al. 1960) developed by regional scientists. The most straightforward example of this cross-fertilization process and its implications for economic geography is provided by Britton (1967) who has employed the concepts and techniques of regional science in an examination of the internal structure and external spatial relations of the Bristol region in south-west England. More generally, Lloyd and Dicken (1972) suggest that as a 'behavioural science concerned with the spatial dimension of economic systems, economic geography is concerned with the construction of general principles and theories that explain the operation of the economic system in space' (2). They, like McDaniel and Eliot-Hurst (1968) and Eliot-Hurst (1972), adopt a systems approach; however, they emphasize the spatial patterns created by the energy flowing through the system and the spatial modifications to the flow of energy rather than the characteristics of the system itself or of the processes operating within it. Lloyd and Dicken structure their study around the nature of the spatial dimension which they simplify in the first part and gradually complicate in the second; the third section combines their considerations of space, both as an economic fact and as a factor, in a study of spatial inequalities in economic development.

The environmental relations of the economy

One of the major points of convergence between economic geography and other social sciences is the increasing concern for the impact of space upon economic development and social opportunity. The real and urgent practical problems of the modern world, more especially in the less-developed world, have stimulated a form of applied geography in which the focus is very clearly upon the identification, analysis and solution of problems of development in a spatial and temporal context. Farmer's (1957) study of pioneer peasant colonization in Ceylon is still a classic work in applied economic geography relating to a developing country and the volumes edited by Ginsburg (1960, 1961) began a whole series of studies where economic development provided the focus of analysis (Mountjoy 1966, 1971; Hodder 1973). Brookfield (1973) has provided a most stimulating review and critique of geographical analyses of economic development, stressing in particular the danger of transferring one set of cultural values to a completely different cultural situation. This danger is increased by adopting techniques of spatial economic analysis developed within more advanced economies.

However, within more developed economies too, public attention is becoming increasingly focussed upon geographical space, and some of the reasons for this have been outlined by Chisholm and Manners (1971). The need for public intervention has also been demonstrated recently by Coates and Rawstron (1971), who analyse the spatial variation of selected social and economic phenomena in Britain. They conclude that many of their maps illustrate spatial variations that are 'unjust, harmful and inefficient' (289). Within the United States and, equally predictably, South Africa there are severe local and regional problems of poverty and inequalities in standards of living (Morrill and Wohlenberg 1972; Smith 1973a and b).

The rapidly growing field of urban economics (Hoch 1969; Thompson 1965; Richardson 1971; Goodall 1972) has helped to focus attention on the problems of social inequality and economic efficiency within urban areas, which form intense and quantitatively important regions of social and economic activity (e.g. Dunning and Morgan 1971). Numerous investigations into the impact of urban spatial structure upon social structure and attitudes (Pahl 1968; Robson 1969), into the changing distribution of real income within urban areas (Pahl 1971; Harvey 1973) and into the problem of decaying central city areas (Eversley 1972a, b and c) form the basis of a growing interest in the social and spatial analysis of urban economies, exemplified most comprehensively by the volume of essays on London edited by Donnison and Eversley (1973).

The contemporary importance of centralized planning for economic development and the significance of socio-economic spatial inequalities in economic planning have been described by Wood (1969), while Prescott (1968) has outlined the major forms of state economic planning. Kasperson and Minghi (1969) and Prescott (1972) have attempted to assess the implications and effects of government policy on society and landscape whilst Linge (1971) provides an approach to the systematic analysis of the economic functions of government. Detailed reviews and assessments of regional economic policy, designed in part to redress spatial inequalities, have been provided in a series of essays focussing upon the policies and problems of individual economic planning regions within the United Kingdom (Manners 1972).

But the understanding of spatial inequalities in advanced economies is by no means fully developed (e.g. House, 1974). The priorities of resource allocation, the specification of clear planning goals and the most appropriate modes of intervention in the economic space transforming processes 'remain more intuitive than clear' (Chisholm and Manners 1971, 20). It is true that decision makers lack much of the necessary spatial data for making decisions (Coates and Rawstron 1971), but a more fundamental reason for the existence and persistence of socially unjust economic inequalities is the widespread acceptance of the socio-economic mechanisms which determine the distribution of economic power within capitalist economies. Harvey (1973) shows that it is these market mechanisms themselves, and not some malfunctioning of the system, that are responsible for social injustice in capitalist society. An alternative set of relationships between economic activity and social norms is contained in the contemporary Chinese vision of society (Buchanan 1970) in which 'human and economic relations are motivated not by considerations of material gain but by moral incentives' (viii). This vision and its implementation provides, according to Buchanan, a greater contribution to the creation of a more human world 'than all the achievements of a technology which is bringing a progressive dehumanization of the society in which it flourishes and whose contribution to alleviating the lot of the damned of the earth is derisory' (viii).

Similar problems of a lack of information and an inappropriate institutional framework surround the relationships between economic processes and their ecological consequences in economies reliant upon high technology. Simmons (1966) has spelt out the need for ecological considerations in economic decision making, and the ecological component of regional science has been developed at a conceptual level by Isard (1968). More practically, O'Riordan (1971) and Burton and Kates

(1965) have examined some of the concepts and techniques involved in the management of natural resources which can provide an important means of conserving the ecosystem in the face of a sustained economic attack, whilst Isard *et al.* (1972) have made a regional case study of what is termed ecologic–economic analysis. But Simmons (1973) points out that 'conservation is about human values – what a cultural group desires from its environment and what it will accept by way of change in order to get it' (271). Clearly, human value systems are as neglected in the geography of ecological control as they are in the geography of social change.

Economic geography and the economy

Whilst it is convenient to identify differing viewpoints in the study of economic geography, it must be recognized that the categories outlined above are not necessarily mutually exclusive and that criticisms of the various viewpoints are common.

Much economic geography is derided on account of its 'gazetteer' nature and because of its implied assumptions about the absolute nature of geographical values, usually defined in physical terms. Regional economic geography is sometimes criticized as being difficult to distinguish from regional geography *per se*, and regional science and the study of economic development by geographers is said to contain material that is more properly the concern of economists, sociologists or ecologists. Again, whereas some viewpoints are largely descriptive, others are mainly theoretical. Furthermore, concentration of effort in economic geography upon locational analysis and the link between process and form (Bunge 1966; Haggett 1965; Harvey 1969) has been criticized by several authors. Manners (1970) has pointed to the 'glaring lacuna in the research achievements of economic and urban geography' (55) suggesting that little attention has been given to spatial variations in productivity. Put another way, this criticism suggests that whilst economic geographers have given much thought to the *causes* of economic location, they have scarcely considered the *consequences*. Similarly, Robson (1969) points out that the attempt 'to isolate a separate set of interests comprising the geometrical aspects of geography is mistaken in so far as the real interest in the existence of a pattern in, say, the distribution of central places, lies not in the existence of the pattern itself, but in the understanding of the movements and circulations which are responsible for the spatial pattern' (33). In other words, locational analysis is not an end in itself but one means to an end – the understanding of process.

Such criticisms of the different viewpoints in economic geography should be expected and indeed welcomed. As already noted, the field of economic geography is vast and imprecise; it is wide-ranging, eclectic, even interdisciplinary as a field of study and can never be strait-jacketed within any narrowly conceived disciplinary framework. While in no way condoning physical determinism or other simple causal explanations of economic activity, it must be emphasized that descriptive material and basic information involving a direct and practical appraisal of the conditions and facts of world commodities and activities may well be as significant and useful as is any sophisticated analysis of general equilibrium in space. All approaches mentioned in these pages and those 'new' approaches which, undoubtedly, will emerge in the future, are important, legitimate and appropriate for the purposes for which they are intended. This point is perhaps worth emphasizing because of a growing obsession with purely spatial matters. For many geographers today economic geography is to be found where the two disciplines of economics and geography overlap: that is, 'in the consideration of the spatial arrangement of economic phenomena' (Warntz 1965, 54). It is obvious that the spatial expression of economic activities may be important in an analysis of temporal process, but it may equally well be largely irrelevant. To abstract the purely spatial element is to falsify as surely as to abstract the purely temporal or any other single element. It is certain that specialization in spatial and locational analysis, as in many other branches of economic geography, is and should be proceeding apace. The only danger is that any one of these increasingly narrow specialisms or disciplines will be set up as the only 'true' economic geography.

This appraisal of alternative approaches to the study of economic geography raises another, more fundamental, issue. Smith (1972) has pointed out that economic geographers have responded to the prevailing values of society in that 'they have been far more concerned with studying the production of goods and the exploitation of resources than the conditions in which people live'. Thus 'trivial aspects' of economic geography receive detailed treatment in regional texts whilst 'critical contemporary social issues are largely ignored' (17–18). Smith cites the example of the section on the United States in the Aldine University Atlas in which there are no maps of personal income levels or social conditions. The spatial distributions of hogs, turkeys and chickens are, however, considered sufficiently important to take up illustrative space.

Our approach in this short introduction to economic geography contains some elements from all the viewpoints already mentioned. But two points should be made here. In the first place, we are concerned to demonstrate the complex but essentially integrated nature of economic activity within economies at widely differing levels and scales; for this reason we emphasize the structure, component parts and interaction between those elements which characterize economies.

It is for this reason too that the scale of analysis is largely micro-economic. We are less concerned with temporal fluctuations in the behaviour of the economy as a whole than with the behaviour of the individual elements and the flows between them.

Secondly, our emphasis is on process and function rather than on form or spatial patterns. Attention is focussed upon the role of government, firms and resource owners as major units of decision making governing the behaviour and organization of economic activities. It is, of course, impossible to avoid the interaction between temporal process and spatial form, but in general we agree with Ginsburg (1969) who suggests that current revolutions in geographical thought are concerned less with quantification than with a movement away from a 'widespread preoccupation with patterns and distributions' towards an 'emphasis on the organization of area and on the social behaviour which underlies that organization' (403). In short, we accept the notion that in contemporary geography our 'primary concern is with organization and behaviour as *processes by which patterns are generated*' (403). Furthermore, we suggest that these 'patterns' are but one consequence of the processes of economic activity, which also have a fundamental influence upon the social and political structuring of society and upon ecological balances within the environment. An overriding emphasis on spatial form not only tends to obscure such issues: it also limits our understanding of the very forces which give rise to spatial patterns and it encourages the development of a 'trivial' economic geography. By basing our approach upon the concept of the economy, we have tried to ensure that this introductory study of economic geography is undertaken in a way which emphasizes the nature of simple economic processes and highlights the economic basis of many social inequalities and ecological imbalances.

Part 2
The economy: decision making

Decision making underpins economic activity. It is undertaken by the elements of an economy but they operate within a given institutional framework embodying many social, cultural and political values. These are crucial influences upon decision-making behaviour but they are not often appreciated by the decision maker or by students of economies. In this section we shall consider decision making in the economy as a whole before a more detailed study of the individual decision-making elements.

3 Decision and control

Economic activity consists of several closely related stages: the generation of wants and needs known collectively as demand; the transmission of information about the nature of demand to firms; the transfer of information from firms to resource owners about the demand for resources; the response of resource owners by mobilizing their resources; the movement of resources to firms in return for income; the manufacture of goods; the transmission of information from firms to consumers about the availability of goods; the movement of goods to consumers in exchange for payment; the consumption of goods by final consumers. All these aspects of economic activity refer to those tangible, often visible, processes which are apparent in everyday life and whose energy creates the economic landscape. But it will already be clear that there is another and perhaps more fundamental aspect of economic activity – decision making – which provides the impetus and direction for these processes.

Three sets of decisions
All economies have somehow to make at least three fundamental and interdependent sets of economic decisions. These are *allocation decisions* – what commodities shall be produced and in what quantities; *production decisions* – how and where these goods shall be produced; and *distribution decisions* – for whom they shall be produced.

ALLOCATION DECISIONS
In a society where economic activity consists of the concerted action of large numbers of individuals, decisions about which goods and services and in what quantities these goods and services are to be produced are exceedingly complex. Individual views on the most effective use of scarce productive resources will vary widely, so that the economy must make value judgments about the kinds of goods to be produced and quantity judgments about how much of each good should be produced. In all economies the criteria of efficiency and utility should govern the decisions because allocative efficiency is measured by the ratio between useful output and total input. The effect of such decisions is to allocate resources to those productive agencies capable of producing the chosen goods.

PRODUCTION DECISIONS

Having decided on the type and amount of goods to be produced, the next set of decisions involves the organization of production. Production is a process involving firms in the acquisition of premises at a suitable location in order to combine resources in an efficient productive ratio and to set in motion the distribution of the finished products. Again, the role of utility modifies any purely physical interpretation of efficiency. It is often not possible to use resources in the occupations, locations and ratios in which they have the highest level of productivity (that is, the highest output–input ratio), because the constraints of the utility priorities of society may well modify this simple relationship. Both resources and demands are distributed unevenly over the surface of the earth. Some communities, therefore, must attempt to compensate for their lack or relative scarcity of certain resources by utilizing their limited resource endowment in ratios which make extensive use of plentiful resources and intensive use of scarce resources. Alternatively they may, by military or economic conquest, extend their access to a wider range of resources or, by developing trading connections, import those goods which are demanded by their consumers but which are difficult, if not impossible, to produce with their own resource endowment. As a result of these considerations, decisions about methods of production need to be made at the same time as decisions about which goods and services are to be produced, because the output of goods depends not only upon utility priorities and resource endowment, but also upon the ways (production techniques) in which the resources can be combined in productive service.

DISTRIBUTION DECISIONS

In a subsistence economy the producer of goods and services not only provides his own factors of production but is also the consumer of the finished product. There is no need for complex arrangements to facilitate resource allocation, production, or distribution of the finished product, as all processes are under the control of one individual or a small group of individuals. However, in a society in which substantial specialization of occupation has developed, each individual produces only a small fraction of his own personal requirements, and in certain occupations (for instance, the manufacture of component parts for machines) the individual may not produce anything of direct use to him or to his family. Clearly, some means of distributing the bundles of goods to satisfy the demands of the members of such a society is necessary. Furthermore, the amount of goods distributed throughout the economy must be equal to the amount of goods that it is possible to produce with the available resources, and the type of goods must bear some relation to the demand priorities of the consumer house-holds. Decisions must, therefore, be made both about the quantity and type of goods each member of the society should receive. These distribution decisions are thus closely related to both allocation and production decisions and must be taken at the same time.

The mechanics of decision making

Decision-making processes operate under various conditions which may be regarded as lying along a continuum. At one end decisions are made by a central authority with power to govern the elements of an economy; the type of economic system resulting from such an organization is known as a command economy. At the other end of the continuum decisions are made entirely by the individual consumers, firms and resource owners without any interference from a central authority; this is known as a capitalist or market economy. In reality, however, few economies nowadays operate in either of these two ways, most economies more realistically being described as 'mixed', in which both private and public decision-making contribute to the end result. Nevertheless it is pedagogically convenient, as well as common practice, to make this distinction between three broad types of economic organization – command, capitalist and mixed.

In a command economy the means of production are collectively owned and the decisions made by a central administration which needs to gather all relevant information together and keep it constantly updated. Three sets of information are required: the size and nature of the demand for goods; the different ways in which resources may be combined in alternative production methods; the size and nature of available resources. With these data to hand, the decision-making process involves the drawing up of a list of priorities for the use of productive resources, choosing efficient methods of production, matching these methods of production with the list of priorities and deciding upon the criteria for distributing goods and services amongst the community. Clearly, in such an economy both the data-collection processes and the decision-making processes are complex, large and exhausting tasks and whilst there are many advantages in a command economy, associated with its ability to extend the range of criteria for decision making, it is hardly surprising that mistakes may be made by the central authority to the detriment of the economy as a whole.

Some of the advantages and disadvantages of this system for economic–geographical decisions are being explored in a small body of literature dealing with the behaviour of decision makers within planned economies. For example, decisions about industrial location in Eastern Europe are, according to official sources, based upon ten principles or objective laws of socialist location (Hamilton 1970). However Hamilton

(1971) has pointed out that the existence of such principles does not remove the need for selecting those that are relevant to particular location decisions, for reaching a compromise between the principles and the economic and technical needs of industry, or for resolving the conflict between economic and social principles. In fact, decision makers cannot work objectively in applying the principles because their behaviour is itself conditioned by several environmental influences. These include the temporally variable strength of their ideological beliefs; a conflict-ridden decision-making system created by pressure groups, the emergence of which is itself associated with changes in the acceptance of Marxist–Leninist philosophy; an economic environment stripped of market pricing which cannot provide any clear rules or data on economic efficiency but which is nevertheless shifting towards economic criteria and away from those of ideology; an environment in which spatial concepts are inadequately researched and distorted by the decision makers' identity with a home area, by various spatially defined pressure groups and by an imperfect supply of data. The tendencies towards industrial dispersion or concentration embodied within the principles of location and observable at different times in Eastern Europe appear to be associated with fluctuations in the decision-making environments, rather than with the objective application of the principles, and so result in a reduced supply of information feedback about locational alternatives and data for locational selection.

Although economic decision making in the Soviet Union is more centralized and ideological in character, evidence suggests that some of the conflicts discussed by Hamilton are faced by Soviet decision makers. During the nineteen-sixties the regional investment of industrial capital within the Soviet Union favoured the less productive eastern regions at the expense of under-capitalized regions in the west (Dienes 1972). This policy derived from the attempt to replace the purely economic criteria of profitability by strategic considerations and from a desire to develop the water power and mineral resources of the Asian hinterland, despite its harsh physical conditions and poorly developed infra-structure. It has, however, produced a conflict situation. The pro-European factions point to the poor return on Siberian investment, the manpower shortage and the difficulty of recruitment to the east and criticize many of the large-scale, energy-based projects in the eastern areas. This conflict between economic and non-economic criteria for decision making may be illustrated by one more example, again taken from the Soviet Union but this time in the field of agriculture (Jensen 1969). An inherent contradiction has arisen out of the desire to place all farms on an equal operational footing and so ensure equal returns to labour, despite variations in physical conditions and, at the same time, to promote agricultural specialization in conformity with natural and economic zones. The problems are made even more difficult because of the lack of adequate and sufficiently detailed data to permit appropriate decisions to be reached. Clearly economic efficiency and social justice are related in a very complex manner and the benefits to be derived from using a wide range of criteria for decision-making must be qualified by the costs of so doing.

In capitalist, and to some extent in mixed economies, the decision-making process is decentralized from government to other elements in the system. Decision making in a decentralized economy is based upon private ownership and gain and is undertaken by households – as consumers and as owners of resources – and by firms. Consumers choose the type and quantity of goods that will yield them the greatest utility. But the extent of their freedom of choice will depend in part upon their knowledge of alternative commodities and upon their spending power which, in turn, is determined by the income they receive as resource owners. Firms, as agencies of production, must decide which lines of production, what techniques of production and what locations of the producing units will yield them the greatest benefit. If we accept the àssumption that firms are motivated largely by economic considerations (Baran and Sweezy 1968, 33–40) this 'benefit' is measured by profit, or the excess of incoming revenue over outgoing costs. Production decisions will affect the amount and type of resources used in the production process and householders, as owners of these resources, will attempt to maximize their incomes by selling their resources to the highest bidder. Again, if we assume that owners of resources are rational in an economic sense, then income will refer to financial gain. However job-satisfaction, a sense of vocation, simple ignorance of alternative uses to which resources may be put, or a weak bargaining position can easily distort this simple objective of income maximization.

Clearly, the point at which the interests of both households and firms coincide is in their consideration of prices. For it is here that a common factor affects consumer utility, the firm's profit, and the resource owner's income. The price of a good, as of a resource, will depend in part upon the amount of goods or resources that firms or resource owners are willing to produce or offer in relation to the demands that are being made upon them. For a price to be mutually satisfactory to both parties the quantities of goods or resources demanded must equal the quantities that are being supplied. If, for example, more resources are being offered than are needed in the production process, then the owners of these resources will accept lower prices for their services. This should cause resource demand to increase as firms can take advantage of low-cost resources by using more of them in the production process. If their revised demands exceed the availability of resources, then firms will compete with each other for the use of these resources by offering higher prices. This should then produce more of the required resources as their owners start to respond to the incentive of higher prices. The price will continue to fluctuate and will not stabilize at the 'equilibrium price' until demand for the resource is equal to the supply. Exactly the same considerations apply to goods.

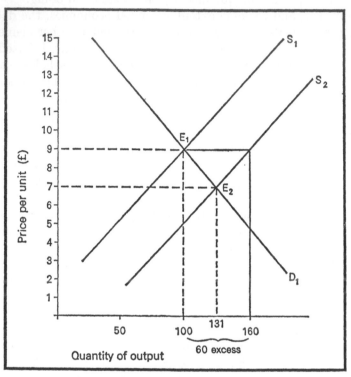

3.1 Demand, supply and price formation.

The above verbal description is represented graphically in fig. 3.1 which attempts to illustrate the relationship between demand, supply and price. For most commodities consumer demand falls as the price increases. This is shown by the demand curve (D_1) sloping down from left to right. Conversely, firms are usually willing to increase their production with increases in price and so the supply curve is shown sloping down from right to left. The point (E_1) at which the supply curve (S_1) and the demand curve (D_1) intersect is the point at which the amount of the good demanded by consumer households is exactly equal to that which firms are willing to supply. One hundred units will be supplied and consumed at a price of £9.00 per unit. This is the equilibrium price. If the conditions of supply change so that firms decide, or are forced to offer, more produce for sale at any given price, the supply curve shifts to the right (S_2). This means that at the old equilibrium price there is an excess of 60 units of supply: more of the commodity is being offered on the market than consumers are willing to purchase at that price. As a result, firms reduce their price and attract more buyers (movement from left to right along D_1) so clearing the market until quantity demanded is equal to quantity supplied at E_2. This is the new equilibrium price under the new conditions. Exactly the same process operates in reverse for conditions of excess demand where demand for goods exceeds the quantity supplied at the old equilibrium price.

The three decisions, described earlier for the command economy and involving the central administration in much work, have now, in the free but unrealistic market economy, been accomplished simultaneously and automatically by the price mechanism operating through the processes of decentralized decision making. The demand conditions for goods have been transmitted to the firms by their price levels. Firms tend to choose to produce those products for which there is an intense demand and to use those methods and locations of production which, given the availability of resources and the location of consumers, will minimize costs and maximize the quantity of goods which can be sold at a given price. In turn, the resource owners will, if they wish to maximize their income, sell to the highest bidder so that resources will be allocated to those firms paying the highest price. A high price for resources can only be paid by efficient firms producing commodities for which demand – and hence price – is high. Thus the problems of allocation and production are solved simultaneously, resources being allocated automatically to those firms producing most efficiently for the most insistent consumer demand. The problem of distribution is also solved because, in attempting to maximize their income, resource owners also maximize their spending power as consumers. Thus they are able to generate an effective demand for goods in direct proportion to the product of the quantity of resources offered to firms and the price at which they were exchanged. Thus the value of resources sold into the economy will determine the power of consumers in the market for goods and services and so solve the problem of distribution.

This latter point underlines the integrated nature of economies. For, as consumer demand increases and firms respond by increasing production, there will be a simultaneous increase (although not necessarily of the same dimensions) in the demand for the requisite factors of production. This will bid up their price. Consequently resource owners, in order to continue to maximize their income, will be tempted to redirect the sale of their resources to those firms offering the higher prices. Thus, as the demand for goods changes, so should the allocation of resources change to feed those firms producing the goods currently demanded, until the point is reached when the need for extra resources within these firms is satisfied. This process can cause severe problems of adjustment in those sectors of the economy facing stagnant or declining demand conditions. The demand for resources within such sectors may also stagnate or decline and cause resource unemployment, unless owners can move their resources to another more profitable sector. Similarly, low levels of profitability within a sector of an economy, or even a complete regional or national economy, can lead to a loss of resources from the sector or area, to a further loss of efficiency and to a decrease in its share of total economic activity.

The assumptions of perfect competition

The description of the price mechanism as a means of making decisions and organizing an economy outlined above, is presented in probabilistic rather than deterministic terms but it relies for its validity upon a number of grossly unrealistic assumptions. Amongst these is the assumption that there is a large number of buyers and a large number of sellers. Under such conditions, if any single buyer increases his demand or a seller increases his supply, the impact on the total demand and supply conditions will be negligible. The importance of this assumption is that no buyer or seller has sufficient economic power to influence price by his own individual action. A second assumption is that all the goods exchanged in the perfectly competitive market are identical; the goods may be distinguished only by price. Thirdly, there is the assumption of freedom of entry into the market by sellers. This works against collusion, among any group of sellers, to push the price beyond its competitive equilibrium. If they do so new sellers will be attracted to the market, the supply will increase and the resultant competition for markets will push the price down. Both buyers and sellers have perfect knowledge about prices in other parts of the markets. This means that no temporary inequalities in price can exist during the time it takes for the news about changed demand and supply conditions to permeate the whole of the market. Finally, there is the assumption that there is no form of distance friction, that the economy operates within a dimensionless point, i.e. that space does not exist, spatial inequalities in price are therefore not considered because they could never arise. Taken together these assumptions produce a market situation operating under conditions of perfect competition. They result in the rapid adjustment of the market to changes in demands and supplies which in turn result in the constant tendency towards an equilibrium price. Furthermore, they artificially remove any tendency for economic power to become concentrated within any single element of the economy.

Price and time

The influences of both time and space cause modifications to this model of perfect competition. In reality, rapid changes in demand cannot be accommodated instantaneously by changes in supply. Resources have to be acquired and prepared and the production process itself may take some time to complete, whilst rapid fluctuations in supply may result from uncontrollable natural causes. The latter is particularly true in the case of agricultural products and the former in the case of complex pieces of technological equipment. As a result the decision to increase supply is difficult to make because, by the time increased supply has been achieved, the demand parameter may have changed causing a reduction in demand so that a condition of excess supply develops with a consequent lowering of price. Under such conditions, supply coming on to the market at any point in time is usually the result of decisions made by firms in the past, whereas decisions made about current production always have their effect on supply at some time in the future. Such a situation induces a fluctuating price level.

Furthermore, individual buyers and sellers may react quite differently to changes in price and such reactions also vary from product to product. Thus a price fall in one commodity may induce an even greater proportionate rise in demand, whilst for another commodity only a slight rise in demand may result. Similarly, different groups of buyers may change their demands to a degree more or less than a given change in price. The relationship between changes in price and changes in demand and supply is known as the price elasticity of demand or supply ($E_{D/S}$), normally given as:

$$E_{D/S} = \frac{\% \text{ change in quantity demanded/supplied}}{\% \text{ change in price}}$$

An elasticity of 1 clearly indicates a one-to-one relationship between the two variables. Values less than one show that demand or supply is inelastic, or unresponsive, to changes in price, whilst an elasticity greater than one shows a price-elastic demand or supply situation. As suggested above, elasticity may be measured for different commodities or groups of buyers and sellers.

Price and spatial structure

By integrating demand and supply conditions the price mechanism provides the elements of an economy with information for making decisions. It helps to govern the relationship between consumers and firms in the market for goods and services, and between firms and resource owners in the market for factors of production. A spatial manifestation of this process is the formation of market supply and demand areas.

If the price of a good or resource is set at the point of production, the consumer or firm must add to this price either the cost of journeying to purchase the product or the cost of having the product delivered. This will mean that the greater the distance separating the buyer from the seller, the higher the price that must be paid. Under such an arrangement, known as f.o.b. (free-on-board) pricing, all buyers are treated equally in the sense that differences in the price that they must pay are due to differences in the cost of supplying the good. This is shown in fig. 3.2, in which two separate firms (L and R) serve the market area around them. The limit of each market area is a zone in which consumers are indifferent as to which supplier is patronized because the price of goods from one is very similar to that from another. The market areas are defined by the intersection of the price curves

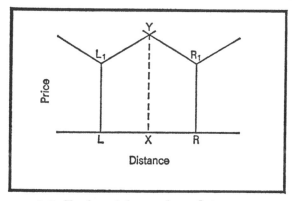

3.2 *F.o.b. pricing and market areas.*

L_1Y and R_1Y at Y. The market area of L lies to the left of X; the market area of R lies to the right of X. But this is not the whole story. Not only do more distant consumers within a market area become less tied to a given supplier, they also consume less of his product (fig. 3.3). The unit price of a good at the point of production (S) is set at a value of p with the result that buyers nearby purchase QS units. With a transport rate of r per unit of distance the price at N is $p + nr$ and the quantity consumed QN. The increase in the price to $p + fr$ at F reduces demand to nil. Thus there is a maximum distance (range) over which the firm can sell its good, and the total quantity of the good sold will be given by a demand cone produced by rotating the triangle FSS_1 around SS_1. If this quantity exceeds the minimum level of demand (threshold) necessary to make the venture commercially successful, then production will continue. The formation of market areas and demand cones involves a certain degree of economic closure and the concept of market areas does indeed provide one technique for defining the spatial extent of economies.

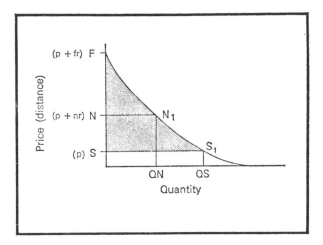

3.3 *Price, distance and spatial variations in demand. (Source: based on Lloyd and Dicken 1972.)*

Space and price formation

Space removes the conditions of perfect competition and so modifies the operation of the price mechanism. This is true even of agricultural products. Warntz (1959) has demonstrated the close relationships between spatial variations in the dependent variable of price and the independent variables of economic population potential (demand) and commodity supply potential (fig. 3.4; table 3.1). Similar relationships hold between price and supply-time potentials, reflecting the seasonal nature of production and the progressive diminution of stores. Thus, in the managed cereals market of the EEC the intervention price, at which support buying begins, is set to its highest level at Duisburg in the Ruhr, and to its lowest level, for soft wheat and barley for example, at Chateauroux in central France. Such spatial distinctions reflect relative levels of availability at the market centres, just as the monthly increments to the intervention price reflect the seasonally changing relationship between demand and supply. Together they are

TABLE 3.1 *Spatial correlations between demand, supply and agricultural prices in the United States*

	Wheat	Potatoes	Onions	Strawberries
Best estimates of multiple correlation	0·87	0·69	0·79	0·69
1% significance level of correlation	0·52	0·53	0·57	0·60

Source: Table XIX (p. 85); Table XX (p. 86), Warntz (1959)

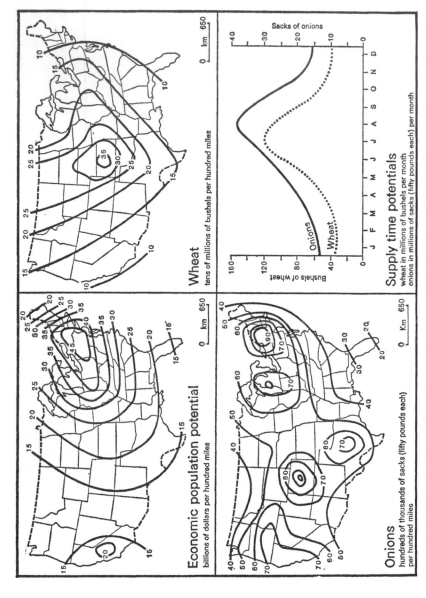

Economic population potential
billions of dollars per hundred miles

0   km   650

Wheat
tens of millions of bushels per hundred miles

0   km   650

Onions
hundreds of thousands of sacks (fifty pounds each)
per hundred miles

0   Km   650

Supply time potentials
wheat in millions of bushels per month
onions in millions of sacks (fifty pounds each) per month

Sacks of onions

Bushels of wheat

Onions

Wheat

J F M A M J J A S O N D

3.4 *Agricultural demand and supply potentials in the United States. (Source: based on Warntz 1959.)*

designed to facilitate the free movement of cereals. But agriculture is a special case. The large number of buyers and sellers and the spatially extensive production systems result in spatially continuous variations in agricultural prices susceptible to analyses, like that of Warntz, based upon the potential concept (Tegsjo and Oberg 1966; Richardson 1969, 38–41). Agricultural commodities, however, account for only a small proportion of total output in developed economies.

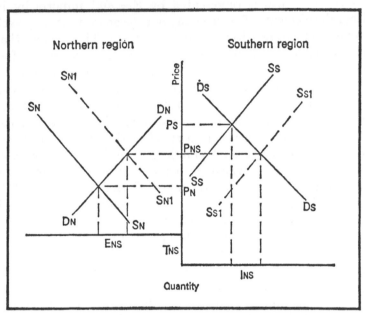

3.5 *Spatial price variations in an urbanized space-economy.*

In industrial, urban societies geographically separate but spatially concentrated economies in large city-regions represent a closer approximation to reality. They introduce space into the price-forming process by replacing the spaceless single market, containing a large number of buyers and sellers, with several markets within which the conditions of perfect competition are approximated but between which the amount of interaction is reduced as a result of the cost of overcoming distance. A simple two-region example is given in fig. 3.5. The supply and demand curves for the production of primary goods in two city-regions, North and South, are shown back to back. The favourable resource endowment of the North, coupled with a relatively low demand, results in a low price (P_N) in the North but this is matched by a high price (P_S) in the South. The horizontal axis of the North is raised by an amount equal to the unit cost of transport of primary goods from North to South (T_{NS}); but despite this, the difference between P_N and P_S is still greater than T_{NS}, so that movement of primary goods takes place. The consequent reduction in supply in the North and increase of supply in the South shifts both the supply curves to the right and so increases equilibrium price in the North and decreases it in the South. The movement continues until the equilibrium price (P_{NS}) is reached at the point where the price of goods in the South equals the price of goods in the North plus transport costs from the North; that is where $P_{NS} = P_N + T_{NS}$. At this point the quantity of exports from North to South is represented by E_{NS} and imports by I_{NS}.

Neither the elements of an economy nor the goods that they exchange can move over space without incurring costs. This means that few buyers can have equal-cost access to a given good offered for sale, and few sellers can have cost-free entry or exit to supply a given market. The cost of transport incurred by both buyers and sellers in reaching a market will increase with the distance over which they or their products have to move. This cost not only inhibits buyers from patronizing a more distant seller, it also restricts the geographical mobility of sellers and their products. A practical example of this sort of spatial influence on price is provided by a study of retail grocery prices in selected towns of Northern Ireland (O'Farrell and Poole 1972). A nearest neighbour analysis of towns with a similar price for a basket of 19 standard grocery items illustrates that there is a strong suggestion of spatial oligopoly in the marked clustering of same price-cell towns in the Belfast Lough and Strongford Lough zones (fig. 3.6a). In this particular case spatial oligopoly describes the small number of shop-entrepreneurs operating within spatially isolated market areas consisting of a town and its hinterland. Under these conditions collusion on price-level fixing between the areas is relatively easy to organize, especially as multiple chains and wholesale buying groups increase

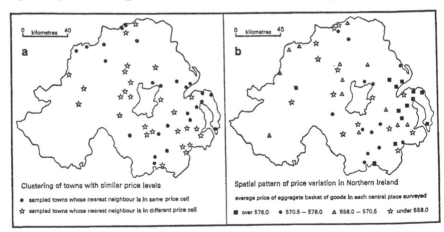

3.6 *Retail grocery prices in Northern Ireland.* (*Source: O'Farrell and Poole 1972.*)

in importance, because of consumer reluctance to incur the costs of travelling to more distant centres for lower prices whilst transport costs prevent more distant sellers from supplying the high-price markets. If prices were fixed very much higher than in more distant towns the long-term solution might well be the movement of shops into the high-priced markets. This would result in an increase in supply and competition and so might help to push prices down. Some evidence for such a hypothesis derives from the finding that price variation between stores within individual towns was less than the price variation between all stores in the area. The forces of collusion operate more effectively over short intra-urban distances than over longer inter-urban distances and, in fact, low incomes depress prices in rural areas (fig. 3.6b).

In this and other ways distance can offer protection from competition to sellers. It may induce them to introduce a discriminatory

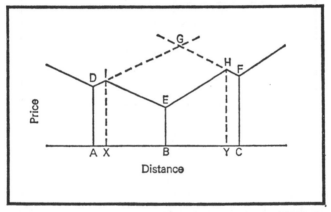

3.7 *Spatial price discrimination. (Source: based on Chisholm 1970.)*

form of pricing policy in which spatial differences do not result from differences in cost of supply but from the degree of economic power conferred upon the seller by space. Thus in fig. 3.7 firm B could charge prices delimited by the line IGH. Such prices would exploit B's protected status to the full, but they might eventually also induce higher cost producers to move to a location close to B in order to gain access to the high-priced market. Furthermore, if B is a tenant, then the existence of high profits in B might cause the owner of the land from which B operates to push up the rent and so reduce profit. Another restriction on B's activities is provided by the reduction in total revenue as a result of cutting off sales by increasing prices; only if demand is price-inelastic could a policy of high prices be effective because, under these circumstances, demand and consumption would fall by less than the proportionate rise in price and so total revenue would increase. But distance itself can induce price inelasticity of demand for goods.

More distant consumers may have a higher demand elasticity than consumers close to the point of sale because the extra costs that would be incurred in reaching, or being reached by, an alternative seller are much lower than those for consumers at the centre of a market area. Consumers at the mid-point between two sellers may be completely indifferent as to which seller they would patronize. By offering a reduction in price to more distant consumers a seller can increase the

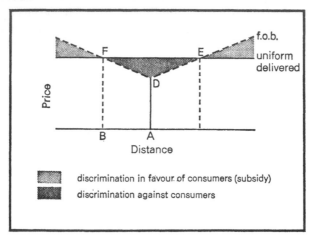

3.8 *Uniform-delivered pricing.* (*Source: based on Lloyd and Dicken 1972.*)

size of the market area over which he can sell his goods whilst the cost of this price-subsidy could be met by increasing the price to nearer consumers whose demands are less elastic because of the excessive distance between them and alternative supply points.

Uniformed-delivered prices, incorporating costs of insurance and freight (c.i.f. pricing) discriminate over space in this way. By imposing a uniform c.i.f. price (fig. 3.8) the firm subsidizes (discriminates in favour of) distant buyers beyond B and C, whilst buyers located close to the supplier are discriminated against because they pay more than they would under f.o.b. pricing. Thus the market area is extended and, provided that the extra sales more than cover the extra costs of transport, such a policy is economically worthwhile, at least from the individual firm's point of view. The advantages of c.i.f. pricing for firms are several (Chisholm 1970, 181–183). They include reductions in invoicing costs and calculating geographical price variations, promotion of sales by facilitating advertising strategy and reducing the complexity of product pricing, reduction of temporal variations in demand by selling to consumers over a wider area and, provided that extra sales are generated, the achievement of scale economies which result from using existing machinery to capacity, or even buying larger more productive production plant.

A final example of spatially discriminatory pricing is basing point pricing (e.g. Rodgers 1952; Melamid 1962; Odell 1975). This is a system which can be used to protect the interests of immobile, long established firms from competition by distant competitors. Under this system all supplies are assumed to come from a single point (the basing point) and so the ex-works price and the costs of transport for all firms in the industry are set from that point regardless of the actual source of supply. In fig. 3.9 the market areas under f.o.b. pricing of the two firms A and B are delimited by F. If A is the basing point then the actual

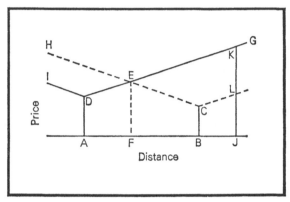

3.9 *Basing point pricing.* (*Source: based on Chisholm 1970.*)

prices to be paid are shown by IDG. A consumer at J pays JK rather than JL. One result of such a system is that all firms will quote identical prices to a buyer – any locational advantage resulting from the proximity of buyer and seller is negated. Cross hauling may also result from peripheral demands, J in fig. 3.9 being served from central supplier A rather than local firm B. Large firms supplying the whole market will tend to congregate around the basing point, whilst smaller firms will move to the periphery and earn an extra profit from the addition of the 'phantom freight' charge (e.g. LK in fig. 3.9) to their profit (Richardson 1969, 38). Some of these effects may be modified by the adoption of multiple basing points which allow greater spatial disaggregation of market areas and so reduce the magnitude of price distortion.

Spatial variation in retail price may reflect the spatial structure of retail outlets as well as the locations of buyers and sellers. However, the Northern Ireland study did not reveal any evidence to suggest that the hierarchy of shopping centres is a systematic influence upon spatially variable prices and an intra-urban study of Swansea (Campbell and Chisholm 1970) showed that type of retail organization was the most important explanatory variable of grocery prices, although the related variables of distance from the town centre and size of outlet were shown to be of secondary importance. A different set of results was obtained in a study of spatial price variation in the central-place system of mid-Wales (Rowley 1972). Four groups of centres were defined in terms of the size of price reduction that they offered for a bundle of grocery goods considered to be frequently subjected to price cutting. The close conformity of this grouping with that based upon the central-place hierarchy led to the conclusion that cut-price grocery goods are, in effect, high-order goods because entrepreneurs in high-order centres are most easily able to generate extra custom from their extensive market areas, to offset the substantial price cuts found in such centres. This opportunity does not present itself to shops in low-order centres and so contributes to the decline of their grocery outlets.

Government intervention

We have seen that the price mechanism is, in theory at least, an effective form of economic control and means of economic integration. We have also seen that governments are important elements within any economy. At the very least they can exert pressure on the other elements by means of their policy-making powers, which may be designed to modify the behavioural environment of the decentralized decision makers, and through their direct control over a large part of national income. They derive this partial control of economies by taxing the elements and their interactions and they expend the money so received on producing and purchasing a variety of public goods and services (e.g. table 3.2). At most the government may take over the running of an economy by assuming all decision-making powers. This does not remove the other elements of the economy, but it places them under the direction of the government.

There are three major reasons for government intervention in the economy. First, the price mechanism may break down and may not perform its task of direction and control in an acceptable manner. Secondly, as we have noted above, the economy has important external relationships: the price mechanism, emphasizing as it does private short-term considerations, is not suitable for controlling the public, long-term interaction of the economy with its environment. Finally, the price mechanism contains within it certain assumptions about the measurement of human values that represent a potential threat to the *status quo*.

Under the conditions of perfect competition it is assumed that the elements of an economy are willing and able to respond instantaneously to purely economic stimuli. For example, factors of production would automatically move from place to place in response to spatial differences in earnings until the forces of demand and supply evened out earnings differentials. But in reality there is a multitude of constraints on the occupational and geographical mobility of resources. For example in the case of labour, social and family ties with a house or a home area, the personal costs of spatial movement and the difficulties imposed by occupational specialization upon the transfer of skills from one job to another may all combine to reduce mobility.

TABLE 3.2 *Public expenditure in the United Kingdom 1965-75* (*£million*)

| | 1965 | | 1975 | | 1965-75 |
	Amount	% of total	Amount	% of total	Percentage increase
Military defence	2105	14·9	5173	9·5	145·7
Transport and industry	2092	14·8	9725	17·8	290·2
Housing and environmental services	1560	11·0	6696	12·3	329·2
Education	1585	11·2	6840	12·6	331·5
National Health Service	1275	9·0	5260	9·7	312·5
Social Security Benefits	2408	17·0	8918	16·4	270·3
Finance and tax collection	168	1·2	662	1·2	294·0
Debt interest	1456	10·3	4513	8·3	210·0
Other	1488	10·5	6678	12·2	348·8
Total	14137	100·0	54465	100·0	285·3

Source: Derived from Central Statistical Office (1976)

Under these conditions the unhindered operation of the price mechanism would lead to continued spatial imbalances in the supply of and demand for resources, which would increase production costs and hold total production down below its maximum potential. Possible solutions to this problem might include government encouragement to firms in the form of information about resource availability in alternative locations, subsidies to help cover the cost of relocation and locational adjustment, or the establishment of government training centres, in both resource–surplus and resource–deficit areas, to encourage occupational mobility. Such policies would modify the behavioural environment of the individual decision makers, thereby supplementing the operation of the price mechanism as a means of decision making and allowing the government some scope to manage the economy.

Similar considerations underlie the government provision of many public goods and investment in infra-structure. Public goods, like national defence, cannot be withheld from any individual consumer without withholding them from all consumers. This makes their provision unattractive to private firms because payment cannot be extracted from every consumer. In a similar fashion the provision of essential infra-structure, or social overhead capital, in the form of, say, communication systems, is open to use by non-paying consumers and benefits other firms and resource owners in ways difficult to monitor. These characteristics, plus the very large quantity of resources required to produce a minimum length of motorway and the long period before a profitable return is obtained, mean that such productive activity is not attractive to private firms and so cannot be left to the operation of the price mechanism. By operating a system of taxation over the whole community, governments are in a sense 'paid' by all households for the goods they produce. But even this situation is not entirely satisfactory because although government-financed commodities, like health and education services, are available to all consumers, there is a tendency for consumers and areas taking full advantage of the facilities to be subsidized by others who do not (e.g. Brown 1972, 61–67).

(ii) Intervention in the allocation process is vital in another respect. The price mechanism helps to govern the interaction between elements, but the criteria used by the price mechanism are of relevance only to the individual elements themselves as they refer to their private gains and losses and do not take account of any external costs and benefits. Thus the simple spatial price differential for resources, which may induce their re-allocation and relocation, is necessarily modified by the costs of congestion in the reception area, by the wasted social investments like schools, hospitals and housing in the source area and by the long-term depressive effect on local incomes and development prospects in the source area caused by the exodus of the most economically active and able resources. These costs derive from, and must be borne by, the economy as a whole rather than one part of it.

Accounting systems based purely upon the price mechanism cannot measure the external, public effect of economic activity. Furthermore, such accounting systems tend to emphasize short-term rather than long-term implications of activity by firms and households, whereas the external implications of economic activity accumulate in the long term. The distinction between private and public costs and benefits and between short and long-term time periods is shown in fig. 3.10. Most private economic decisions are taken using data from the top right-hand box, so that the responsibility of governments should be to consider evidence from the bottom left-hand corner and to modify the private decisions accordingly. This modification is deemed to be so important in certain industries of vital economic and strategic importance to a nation state, that the government takes them over completely. In such nationalized industries, short term, private considerations should not be allowed to outweigh their long-term, public importance.

(iii) The price mechanism is deficient in another, much more fundamental way. As we have discovered, an economy is integrated by the connections between the market for goods and the market for factors

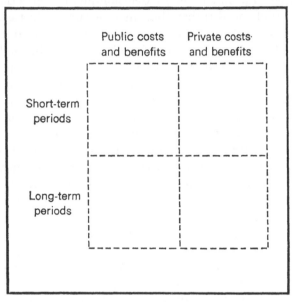

3.10 *Costs, benefits and time.* (*Source: Chisholm 1971b.*)

of production. The income of a resource owner, which plays a major part in his economic effectiveness as a consumer, depends upon the price paid for his resources which in turn depends upon consumer behaviour in the market for goods and services, the behaviour of firms in the production process and the total available supply of resources. In short, the price mechanism in capitalist economies governs the distribution of income (chapter 6) and so reduces the human owners of resources to mere economic automatons. Price can take account of economically effective demands and supplies, but not of human needs and creative capabilities. Furthermore, high incomes allow their recipients to extend and diversify their resource ownership by such means as further education, investment and the purchase of land. In this way they can acquire a larger quantity and more diverse range of resources which may then be sold into the economy. The owner of a large range of resources is also, by virtue of his high income, a highly effective consumer (chapter 4). His demands are backed up by spending power and so are more effective than are those of a low-income consumer without the economic wherewithal to consume. If such a situation of inequality induced by the operation of the price mechanism is not acceptable to a community it is faced with two possible solutions. It may modify the price mechanism by a system of income redistribution favouring need (e.g. Table 3.2) rather than economic worth defined by the market price mechanism, or it may remove the price mechanism, based upon the private ownership of property, and substitute a system based jointly upon collective ownership and a set of decision-making criteria concerned with the values of social rather than mere economic justice.

The integrated nature of any economy ensures that government intervention at whatever point has implications far beyond any single objective. This situation complicates government decision making but it can be politically expedient. A regional policy, apparently designed to reduce socially unjust spatial inequalities in economic health, can also help to assuage demands for greater political autonomy in the region, increase the use of an economy's productive resources, reduce inflationary pressure generated in the more prosperous parts of the country, and improve upon the efficiency of the price mechanism in the location of economic activity. All this may be achieved without incurring a substantial re-allocation of economic power from one group of elements to another within an economy. But government policy may well contain inherent contradictions. To return to the example of resource allocation discussed above, a policy which sets out to reduce spatial inequalities in the returns to labour may also reduce the efficiency of an economy by subsidizing activity in high-cost locations. This in turn may well lead to an overall drop in the demand for resources, as the high costs are passed on to the consumer in the form of higher prices and taxes, or reduce the supply-reliability of production from, say, the agricultural sector which is closely influenced by fluctuations in the physical environment. Thus capitalist economies, based upon private gain, generate conflicts between, on the one hand, economic efficiency, reliability of supply and overall demand for resources and, on the other, the promotion of income equality.

Conclusions

In any economy certain fundamental decisions must be made. The more complex the economy and the more specialized its individual elements, the more difficult it becomes to organize decision making. The price mechanism provides one way of overcoming this problem, but it has many shortcomings, the most important being that it tends to respond to the actions of the most economically powerful rather than the most needy. The cumbersome and labour-intensive administrative decision-making process associated with command economies is perhaps equally open to abuse, but it does have the advantage of being able to incorporate a wide variety of economic, social and political criteria in decision making. The intermediate solution of a mixed economy is regarded with suspicion by those (e.g. Miliband, 1969) who doubt the arbitral ability of capitalist governments concerned with the maintenance of mal-distributed economic power. In any case, the ability of governments in economic management is by no means proven and the priorities of resource allocation change with the evolution or revolution of political ideology. However, we must now leave this vast and complex topic around which so much social and economic history has been constructed and return to our original simplified premise that all economies, developed or less-developed, urban or rural, industrial or agricultural, command or capitalist, are comprised of the same essential decision-making elements of consumers, firms and resource owners. It is to their decision-making roles, whether undertaken for themselves or by a government, that we now turn.

4 Consumers

Consumers engage in the process of consumption: they use up goods and services and so generate satisfaction or utility. This is the opposite of the production process undertaken by firms (producers) employing inputs to yield goods and services. The consumption of resources and the intake of semi-finished goods by firms is part of the process of production and is not explicitly considered in this chapter. We are concerned here with the ultimate use of goods and services by individuals or households to satisfy their demands.

Consumer behaviour and the economy
The behaviour of consumers, as with that of firms, resource owners and governments, is both an input to and an output from the operation of economies. As an input, consumption is important for two major sets of reasons. First, consumer behaviour is an important explanatory variable of the size and temporal fluctuations of flows through an economy. In the United Kingdom economy, for example, the largest part of G.N.P. is made up of consumer expenditure (table 4.1) so that even small fluctuations can cause substantial production adjustments by firms, losing or gaining confidence in their ability to make profits, and so bringing about changes in the level of demand for resources. Attempts to explain the aggregate behaviour of consumers have focussed attention mainly upon their choice between consumption and saving at different levels of income and so upon the aggregate demand for consumer and investment goods produced by firms. Secondly, the behaviour of consumers may be an important determinant of the allocation of resources to alternative uses. Changes in the composition of consumer demand can lead to fundamental changes in the occupational and spatial structure of an economy as new products become popular and traditional goods are gradually discarded; new industries expand and traditional industries decline (table 4.1). The locational preferences and factor demands of the new industries may differ from the old and so resources and space are evaluated anew as locational choices are made and inputs employed to begin the production process.

The relative strength of consumers' influence upon the working of the economy depends upon the extent of consumers' sovereignty. If resources are allocated to those sectors of the economy that produce the goods most desired by consumers (i.e. in line with consumers' preference) then consumers' sovereignty is said to exist. In such a case the operation of the economy is directed towards the satisfaction of the effective demands of its consumers. Consumers' sovereignty may be limited by government action to suppress the influence of consumers' preference,

TABLE 4.1 *Consumers' expenditure in the United Kingdom 1965–75*
(£ *million at current prices, including sales taxes and imports*)

| | 1965 | | 1975 | | 1965–75 |
	Amount	*% of total*	*Amount*	*% of total*	*Percentage increase*
Food	5059	22·1	12092	19·1	139·0
Alcoholic drink	1499	6·6	4902	7·7	227·0
Tobacco	1428	6·2	2741	4·3	91·9
Housing	2592	11·3	9201	14·5	255·0
Fuel and light	1087	4·8	2927	4·6	169·3
Clothing	2099	9·2	5320	8·4	153·4
Durable goods	1785	7·8	4858	7·7	172·2
of which motor vehicles	*707*	*3·1*	*1932*	*3·0*	*173·3*
Travel	1681	7·4	5929	9·4	252·7
by private motor vehicle	*940*	*4·1*	*3940*	*6·2*	*319·1*
by all other modes	*741*	*3·2*	*1989*	*3·1*	*168·4*
Services	3292	14·4	9264	14·6	181·4
Expenditure abroad	390	1·7	1089	1·7	179·2
Other	1933	8·5	5050	8·0	161·2
Total	22845	100·0	63373	100·0	177·4
Total domestic expenditure	35886		105441		193·8
G.N.P.	31656		94095		197·2
Consumers' expenditure as a percentage of: Total domestic expenditure	63·7		60·1		
G.N.P.	72·2		67·3		

Source: Derived from Central Statistical Office (1976)

to restructure or reduce it (table 4.1) in order to establish a different set of influences upon resource allocation. Resources freed from the production of certain consumer goods may be transferred to the production of capital goods or to an alternative set of goods and services desired, but not effectively demanded, by certain groups of consumers. The sovereignty of consumers may also be limited if their bargaining power is reduced *vis à vis* firms in the economy. If the production of certain goods becomes concentrated within a few large firms, consumers are forced to rely upon a small number of supply sources and so are less able to use their power to switch custom from one firm to another in order to exert pressure upon suppliers. As we have seen already, the costs of overcoming distance to patronize alternative sellers may restrict consumers' choice and sovereignty and thereby increase the bargaining power of firms within market areas.

The analysis of consumer behaviour

Consumption and the subsequent disposal of its waste products are the end results of a complex web of consumer behaviour which includes the generation of wants and needs, the translation of these desires into effective demand (the ability and willingness to pay for a good) and the acquisition of the good. It is hardly surprising then that the study of consumer behaviour has attracted students from a variety of academic disciplines including economics, sociology and psychology. In addition, the importance of consumer behaviour for the profits of private industry within capitalist economies has led to the growth of market research, a major growth industry in its own right, concerned with the attempt to predict and so influence the behaviour of consumers before they reveal their preferences in the market for goods and services.

Economic analyses of consumption are not interested in behaviour itself. Their aim is to establish quantified relationships between the volume and composition of goods consumed and certain independent variables, the specification of which is itself a function of the type of analysis being undertaken. Studies of consumption may be concerned with cross-sectional or time-series data, they may analyse individual consumer households or groups of consumer households and they may be designed to analyse total consumption or the consumption of well-defined, individual goods. Time-series analyses of individual commodities emphasize price as a major independent variable whereas, when total consumption is the dependent variable, the effect of different levels of income upon consumption expenditure is of greater interest. Cross-sectional analyses of individual commodities also usually ignore price, if its variation is considered to be insignificant, and income is again analysed as an independent variable. However, in the cross-sectional study of individual commodities the price of substitutes may be included, at least if they are considered significantly variable. Differences in consumer behaviour between households or groups of households may be ascribable to differences in income or household size, as this is normally positively correlated with household income. The differences not explained by income or household size may well result from non-monetary variables associated with household composition, for example age and sex structure and stage in the life-cycle, or from social status and educational levels. These non-monetary variables underlie the hybrid concept of taste, a variable frequently used in economic analyses of consumption to explain residual differences in household behaviour. Whilst it is important to understand the operation of these independent variables for each type of consumption analysis, it is also vital to remember that results drawn from one class of study are not necessarily transferable to another. For example, significant variations in personal income may provide an effective explanation of differences in the consumer behaviour of individuals within a group, but if the concern is with variations in consumer behaviour between groups, and each group has a similar distribution of income, then the power of income as an independent explanatory variable may be considerably reduced.

Unlike economic analyses, social and psychological studies of consumer behaviour are less interested in the attempt to derive a quantified explanation of revealed acts of consumption, than in the reasons why certain independent variables should be relevant to the study of consumer behaviour (Poole 1968). Explanatory variables based upon the influence of family size and stage in the life-cycle; upon socio-economic and cultural status and their implications for patterns of norms, values, attitudes and sanctions; and upon the speed of innovation diffusion about the availability and acceptability of goods have all been emphasized by sociologists. At the level of the individual, psychologists have explored the bases of consumer decision-making even further. Studies of motivation generated jointly by physiological need and attitude formation, and of cognitive learning processes which combine to induce temporal changes in consumer behaviour, are a far cry from the more superficial economic approach to consumption. However, one reviewer of the range and significance of behavioural concepts for the construction of theory in human geography (Harvey 1967) concludes that the level of understanding and control that behavioural science provides over decision-making processes is not sufficient to develop a powerful synthesis of behavioural postulates and spatial patterns.

This categorization of approaches to consumer behaviour is misleading if it implies that the factors emphasized by each discipline are, in practice, independent of each other. The objective world of economic quantities is in fact perceived through social and personal filters before a consumer makes a decision or takes action. As a result the response of several (groups of) consumers to similar economic quantities may be quite different. Furthermore, social and personal filters are closely affected by the economic quantities faced by any individual or group in society. In a capitalist economy the consumer can express his preferences for consumption only if he possesses the spending power to make his demand effective. This is a major reason for the incorporation of income as a variable in the economic analysis of consumer behaviour. Income is derived from the sale of resources into the economy. Thus the size and variability of household income reflects the amount, quality and scarcity of resources that households can offer. But the work situation at which this sale is accomplished has a more fundamental role to play in the study of human behaviour. Households take their position in the social structure on the basis of the rewards of power, wealth, status and prestige allocated to them by the economic evaluation of their resources, particularly of those resources belonging to the major wage earner. Inequalities of reward produced by this process of evaluation are reflected by inequalities in consumer behaviour which consists of two component parts: the opportunity, or possibility to behave in a certain way and preference, or the revealed act of behaviour. In making this distinction Eyles (1971) points out that preference is limited by opportunity. If a consumer or group of consumers does not have the opportunity to express preference (e.g. demands are ineffective in a market economy) then it cannot be revealed. The resultant repressed preferences are those courses of action perceived by certain consumers as impossible to attain. Even more fundamental may be the limiting effects of lack of knowledge about consumer choice which narrows down the range of opportunities, not because they are physically unavailable but because the potential consumer is barred, by ignorance, from perceiving them. This lack of consumer choice is an important contributor to spatial and personal economic inequality and its extent and causes need to be understood as part of any economically egalitarian policies.

But position in the social structure is also an important determinant of position in the spatial structure (Pahl 1968). The most obvious example of such a connection is provided by the competition between households for residential space in and around centres of employment. Such studies (e.g. Evans 1973), with those of consumer travel behaviour derived from urban land-use transportation studies (e.g. Blunden 1971), are the most intensively developed branches of the spatial analysis of consumption. High incomes and status imply an effective demand for space and a high degree of personal mobility; low incomes imply the opposite. High-income consumers are able to compete effectively for accessible and/or attractive locations, well-served by local facilities and access to transport systems. Alternatively they can overcome poorly served locations by virtue of their personal mobility, which enables them to reach out to better served and more competitively priced supply centres, and by their economic power to attract facilities. By contrast, low-income consumers are economically uncompetitive and much less able to improve their residential locality. As a result, a given unit of household income is worth more in a high-income area than a low-income area because of the increased choice endowed by the superior low-cost access to a wide range of facilities and transport. Under these circumstances the real income of high-income earners may be increased and so the structure of residential space regressively reinforces the economic inequalities induced by the reward-giving mechanism of the market for factors of production. Furthermore, if the less affluent consumers account for a lower proportion of total consumption than their numbers alone might suggest (i.e. if their demands are relatively ineffective), then their particular needs are economically insignificant in proportion to their numbers, because they can have only a limited influence upon the profitability of the production of goods by firms and the sale of goods in market places. Effective demand is a criterion by which firms make their production decisions. Ineffective demand is not. In this way the choice available to the less affluent consumers is limited, not only by their income levels, but also by a range of available goods and outlets geared to the more effective demands of affluent consumers. It is for these reasons that the distribution of income over a group of consumers as well as the size of the group must be taken into account in the analysis of their behaviour. Both the total volume of consumption and its qualitative composition may be expected to vary as the distribution of income changes and so induces change in expenditure and saving.

Economic geography and consumer behaviour

Economic-geographical studies of consumption have been overshadowed by a greater concern for the study of production. The limited amount of detailed spatial data and the fact that most of the world's consumers have little or no consumption choices to make, being primarily concerned with subsistence production, may account for this lack of empirical studies (McCarty and Lindberg 1966). But the deficiency also extends to economic location theory which is almost exclusively concerned with the spatial organization of production (Poole 1968) and makes highly simplified assumptions or deductions about the location of demand. This lacuna in geographical analysis is much more apparent in the study of spatial variations in consumer behaviour, measured in terms of the intake of goods, than in the complementary study of consumer behaviour in space[1] whilst acquiring goods although, in practice, these apparently distinct approaches to spatial analysis are interrelated. Both are partially explicable in terms of the other.

[1] Not to be confused with the rather different interpretation of these terms by Rushton (1969).

Acquisition behaviour is conditioned by the type of goods being sought, whilst the intake of goods is limited by the range of accessible shopping centres. However the two branches of consumption geography remain mutually isolated (but see Andrews 1971 and 1973) although two generalizations may be made about both approaches. First they are concerned, like all studies of consumption, with two major groups of problems: the empirical description and analysis of revealed behaviour, and the understanding of the complex process of decision making underpinning the revealed behaviour. In terms of the volume of research output the first group of problems has been investigated by geographers much more intensively than has the second. Secondly, emphasis upon effective demand in the study of consumer behaviour is reflected in the fact that, with some exceptions, both sets of problems have been conceived largely within a commercial context. The consumption of consumer goods and services has received most attention from geographers despite the fact that the consumption of public goods and services plays an important part in their effectiveness as a means of income redistribution. One of the major difficulties facing such non-commercial studies of consumption is the measure of demand or rather its replacement–need. Need is an extraordinarily difficult concept to measure, especially in a spatial context, and attempts to do so (e.g. Davies 1968) underline the administrative convenience of the price mechanism as a means of coping with consumer requirements. But it also underlines one of the paradoxes inherent in capitalist economies – that repressed preferences are generated, and then ignored, by the price mechanism – so promoting social injustice.

SPATIAL VARIATION IN CONSUMPTION

A prediction that studies in the geography of consumption would be devoted extensively to the basic necessities of food, clothing and shelter (McCarty and Lindberg 1966) had already proved to be correct in the concern for the geography of diet (Sorre 1962; Gregor 1963). But a truly economic-geographical analysis of spatial variations in consumption came somewhat later in an exploratory case-study of the geography of domestic electricity consumption in rural Ireland (Poole 1968, 1970). This study is cross-sectional (for the financial year, 1965–6) and proceeds at two levels of aggregation – a micro-level analysis of a cluster sample of 727 individual consumer households, details of which were obtained from a questionnaire survey, and a macro-level analysis of 78 spatial units for which documentary data were utilized. Multiple regression techniques were used as the main analytical tool and they revealed that at the micro-level, income, especially the income of the head of household, was by far the most important explanatory variable and that social class, household composition and size, the period connected to the main electricity supply (a variable affecting the number of appliances possessed) and the availability of alternative fuel in the form of peat were far less powerful variables. As we have already suggested (p. 53), results obtained at one level of analysis do not necessarily hold at another and a similar set of variables analysed at the macro-level showed that the availability of peat (a surrogate variable for the 'price' of alternative fuels) pushed income into second place in the ranking of the explanatory power of the independent variables. The effect of the other variables was minimal. This reversal of rankings is related to the greater variation of peat availability between, rather than within, spatial units and to the large income variation between households within each unit, the spatial variation of which was limited because the positively skewed distribution of income was characteristic of all such units. The three variable relationships between expenditure (independent variable), income and bog availability resulted in a correlation coefficient of 0·4316 at the micro-level and 0·7098 at the macro-level. The suggested reason for this difference is that the effect of idiosyncratic individual behaviour is filtered out at the higher level of aggregation.

Despite this example of the effect of personal factors upon consumer behaviour, the high levels of statistical explanation achieved by adopting an econometric approach to this study underline the importance of economic variables in the study of consumption. But they also point to the desirability of extending geographical studies to incorporate behavioural postulates in time-series analyses of a wide range of goods. Data are still lacking, although sources like the Family Expenditure Survey (Department of Employment and Productivity, annual) provide an as yet neglected source of data (Hecock and Rooney 1968).

A description of spatial variations in the opportunity to consume public services like the National Health Service and education provided by central and local government (Coates and Rawstron 1971) points to the need for making an assessment of all such spatial consumption inequalities, especially as such goods are supplied by governments, and not private firms, in order to try to avoid unjust and harmful inequalities. Descriptions of this kind are capable of extension to statistical analyses of the determinants of regional variations in, for example, the consumption of education. The study of parental attitudes to their children's education in Sunderland (Robson 1969) is an example of a behavioural approach to spatial inequalities in consumption and is designed to isolate locational influences upon attitude formation. Clearly this study is of less interest to the economic study of consumption than it is for urban social geography but, like Poole's work, it does reveal that the spatial dimension has an important influence upon processes which have so far been analysed in a predominantly aspatial context.

CONSUMER BEHAVIOUR IN SPACE

Over a given period of time the observable behaviour of consumers in acquiring goods and services consists of a series of repetitive and less repetitive journeys between their residential locations and locations able to supply the required goods. But such journeys are only part of a large number of movements undertaken by households in the course of their daily life and it is a misleading oversimplification to assume that at any one time human beings are acting solely as consumers, commuters or leisure seekers (Hägerstrand 1970). In fact consumer movements are usually combined with complex multi-purpose trips (Marble 1966; Wheeler 1972) so that the study of consumer behaviour cannot and should not be isolated from the other forms of movement behaviour.

Despite such theoretically desirable indivisibilities, consuming man and the journey to consume have received considerable attention from geographers whose main interest in the revealed behaviour of consumers has been with descriptions and explanations of shopping behaviour in a system of central places. In defining the concept of the range of a good as the farthest distance that a population of consumers is prepared to travel in order to buy the good offered for sale at a central place, Christaller (transl. Baskin 1966) suggested that four sets of factors affect consumer travel:

(i) The size and importance of a central place measured in terms of its centrality or the degree to which it performs central functions. Places with central functions serving a larger region are called central places of a higher order; those serving local areas are of a lower order. Such attributes of central places are important in determining the range of a good because larger centres allow economies in production and distribution and hence in unit cost and price. Furthermore several goods can be obtained on one trip to a higher-order centre – an arrangement which economizes on the cost of shopping travel and so reduces the price to the consumer of the goods purchased. This, in turn, extends the range of the higher-order centre.

(ii) The price willingness of the consumer to purchase certain goods. This is a factor not only affected by income but, as we have already suggested, by the social, professional and cultural attributes of consumers and by their customs and special demands. Price willingness is also closely affected by the consumers' assessment of the opportunity cost of travel.

(iii) The consumers' subjective evaluation of the cost of distance.

(iv) The order of the goods sought (theoretically higher-order goods are offered only in higher-order places), their elasticities of demand and supply and their resultant price at the central place.

The proliferation of geographical studies of shopping behaviour (Berry and Pred 1965) that attempt to modify or confirm this list of factors and to build models of consumer behaviour may be grouped into three broad but distinct classes (Garner 1970). The first group is essentially cross-sectional in approach and is concerned with describing the distance and orientation of consumer travel. Variations in such dependent variables are related, not necessarily in quantitative terms, to independent variables like the income of consumers (Holly and Wheeler 1972) and their socio-economic status measured, for example, by type of house occupied (Nader 1969). A more effective demand for a greater variety and quantity of goods and a higher degree of personal mobility enable higher-income consumers to by-pass local, low-order shopping centres to which low-income consumers are tied, and to patronize more distant higher-order centres even for the purchase of day-to-day convenience or shopping goods. An example of this effect is provided by Davies (1969) in a comparative study of two suburban areas of Leeds. Higher-income consumers shop with greater frequency, favour private specialized outlets and travel out of their local area more frequently than the low-income consumers. Developing this theme, a study of shopping behaviour in several areas of south-east England showed that the physical attractiveness of shopping centres is a significant factor in attracting middle-class custom and in distinguishing middle-class behaviour from that of the rest of the consumer population (Schiller 1972).

Investigations of the relationships between shopping behaviour and the type of good being sought have shown that, as expected, higher-order goods generate more long-distance but infrequent movements than do lower-order goods, but that the relationships are not clear cut, being complicated by factors such as leisure shopping and shopping on the journey to and from work (Ambrose 1968). Thorpe and Nader (1967) have shown that a subjective four-rank hierarchy of shopping centres in North Durham based on an index of centrality was justified by the hierarchical structure of consumer behaviour within the system; but a similar study of the relationship between a functionally defined hierararchy and consumer behaviour in Iowa (Golledge, Rushton and Clark 1966) reached the opposite conclusion. The authors suggest that a description of consumer behaviour deduced from the existence of a shopping-centre hierarchy is not confirmed by the direct observation of consumer behaviour, a conclusion also reached by Day (1973) in a study of behaviour within the hierarchically planned network of shops in Crawley.

The second group of studies on the geography of consumer behaviour in space differs from the first not in the topics of study but in the research design. They attempt to test, using quantitative techniques of analysis, certain specific hypotheses about consumer behaviour often derived from Christaller's (1966) original formulations. For example, in an early and subsequently influential study, Berry, Barnum and Tennant (1962) attempted to explain the distance travelled by rural consumers in south-western Iowa to acquire a variety of higher- and lower-order goods and services. Using regression analyses they showed, for example, that an independent variable measuring the number of central functions in the centre of first choice was significant for all items and that the percentage of distance travelled over paved roads to reach the centres was significant for most items. But negative regression coefficients illustrated the frictional effect of frequency upon distance travelled for low-order goods. This method of analysis has been extended by Murdie (1965) in an investigation of culturally induced differences in behaviour between Old Order Mennonites and 'modern' Canadians living in south-western Ontario when shopping for traditional goods like auto-repair, food, clothing and shoes. By contrast, the shopping behaviour of the two groups was very similar for modern goods such as doctors, dentists, banks, and household appliances, all of which were developed after the initial establishment of the Mennonite culture in the area and were subsequently adopted by them. A similar but less quantitative analysis of cultural differences in shopping between the French and English Canadians of eastern Ontario (Ray 1967) also revealed significant divergencies of behaviour for intermediate order goods.

The concept of range has been carefully analysed by Clark (1968) in a study of shopping behaviour in the urbanized area of Christchurch, New Zealand. Using information about the sources of three convenience goods and three services, collected from a 1% sample of the population, Clark found that in the case of meat purchases only 46·8% of the population patronized their closest centre. In fact the behavioural hypothesis, ostensibly derived from central place theory, that a consumer will patronize the nearest centre offering the good was shown to be applicable only to vegetable and grocery expenditure at low-order centres. Even in these cases the behaviour of 17% and 27% respectively of consumers was not correctly predicted, despite significant differences between the mean distance to the nearest centre and the centre actually used. Christaller's (1966) suggestion that all goods offered at high-order centres would have a greater range than those at lower-order centres was confirmed by this study which showed that high-order centres attracted consumers of low-order goods from greater distances than low-order centres. Evidence on the consumption of services was less conclusive although the distance travelled to the central business district of Christchurch was significantly different from distances travelled to centres lower down the intra-urban hierarchy. Theoretical justification for such behaviour is provided by Bacon (1971) who shows how economies in travel costs can be achieved by substituting less frequent, high-order trips for more frequent low-order trips under the assumption that the aim of the consumer is to minimize total transport costs.

An outgrowth of the unsatisfactory performance of the nearest centre hypothesis is a series of studies of consumer behaviour which attempt to replace it by an alternative based upon an index of shopping-centre attractiveness. In deriving this index its authors (Rushton, Golledge and Clark 1967; Rushton 1969; Clark and Rushton 1970) argue that the use of regression analysis to investigate consumer behaviour is limited in value because relationships established may also hold for the shopping centres that consumers may choose not to patronize. Such analyses measure the spatial preference pattern of consumers, not their spatial opportunities and so can provide only a partial description of behaviour. The attractiveness index attempts to overcome this problem and at the same time incorporate the influence of the size of shopping centres and their distance from consumers. The index (I_{ij}) may be described as:

$$I_{ij} = \frac{A_{ij}}{P_{ij}}$$

where i = shopping centre size classes
j = distance group from household to centres
A_{ij} = number of consumers choosing to patronize a centre in the ith size class and the jth distance group (a measure of preference)
P_{ij} = number of consumers whose location would allow them to patronize a centre of the ith size class and the jth distance group (a measure of opportunity)

This index replaces the nearest centre hypothesis and is used to predict the shopping centres at which maximum grocery purchases are made. It is hypothesized that consumers will choose the centre with the highest attractiveness index. The application of this test to rural households in Iowa (Rushton, Golledge and Clark 1967) showed an improvement over the nearest centre hypothesis, but an intra-urban study of Christchurch, New Zealand (Clark and Rushton 1970) was less impressive. One reason is that opportunity was defined as spatial opportunity. The role of effective demand was not discussed and so repressed preferences were ignored.

Attempts to construct models of consumer behaviour form the last category of geographical studies of consumer behaviour in space. Several types of model have been recognized (Garner 1970; Bacon 1971) but the most frequently used are based upon empirically derived notions of spatial interaction. The basic premise of such models is that movement between an area of demand (e.g. a residential location) and an area of supply (e.g. a shopping centre) takes place in direct proportion to the attractiveness of the supply points and the strength of demand, and in inverse proportion to the distance, actual or perceived, between them. The most widely used shopping model of this kind (Davies 1973) is that developed by Lakshmanan and Hansen (1965) in an evaluative study of alternative location strategies for satellite shopping centres in metropolitan Baltimore. Their model takes the form:

$$S_{ij} = E_i \; \frac{F_j/(d_{ij})^n}{\sum\limits_{k=1}^{n}(F_k/(d_{ik})^n)}$$

where S_{ij} = sales at any centre, j, generated from any zone, i.
             E_i = total consumer expenditure of population in zone i.
    F_j and F_k = attractiveness of centres j and k.
    d_{ij}, d_{ik} = interaction deterrence functions between i and j, i and k.
             n = an empirically derived exponent.

The operation of such models involves a division of the area under study into residence zones. The model then states that expected sales in centre j from zone i are directly proportional to the attractiveness of the centre and the total expenditure available in i, and inversely proportional to the deterrence on interaction between the zone and centre and the competitive attractions of all other centres in the model.

Although gravity and other models of consumer behaviour have been widely used in the locational and size planning of new shopping centres, and although analyses of revealed consumer behaviour provide a partial statistical confirmation of hypotheses that they have set out to test, it is clear that they contribute little to an understanding of the decision-making behaviour of consumers. Two recent reviews of shopping behaviour (Davies 1973; Garner 1970) have pointed to the need for, and the embryonic development of, alternative behavioural approaches in which the emphasis is placed upon the decision-making processes underpinning the revealed behaviour of consumers. However, contributions in this field have been suggestive rather than substantive (e.g. Garrison et al. 1959; Marble 1966). Suggestions include a conceptual model of the variety of factors underlying consumer behaviour and an analysis of their interdependence – revealing age to be the most significant variable (Huff 1960); explorations of the spatial applicability of analytic models, developed by psychologists, of the way in which decisions are made (Hudson 1971); spatial equilibrium models derived from the concept of learning and the emergence of habitual patterns of behaviour (Golledge 1970) and a strategy for research based upon the perception or images of physical space held by consumers and the relationships between these images, the socio-psychological character-istics of the consumer, the nature of the retail structure and the way in which these images are shaped by experience (Garner 1970).

Conclusions

The belief that the spatial availability of goods is not a significant determinant of their consumption (Rushton 1969, 393) seems, intui-tively, to be wrong. However, the examples of the two spatial approaches to consumer behaviour outlined above show, if nothing else, that the geographical understanding of consumer behaviour is poorly developed and that there is a long way to go before our intuitive belief can be tested in any comprehensive geography of consumption which attempts to combine both approaches. Indeed we turn next to the study of a topic, the behaviour of firms, which has suffered as a result of the number of inadequately tested assumptions about consumer behaviour fed into its analysis. The reactions of consumers, which are, admittedly, rarely independent of the influence of firms, nevertheless provide a vital ingredient for their study. The market for goods and services has rarely, if ever, been a place in which equally powerful buyers and sellers could communicate and it now operates as a means by which the persuasive ability of large firms shapes consumer preference (Galbraith 1967). Clearly the diverse influences upon consumer decision making need to be probed further in the interests of any attempt to understand the decision-making behaviour of firms and to redistribute economic power between firms and consumers. It is, after all the maldistribution of this power which is the fundamental cause of repressed preferences.

5 Firms

In theory the demands of consumers beget production by firms, but in practice this implied equality of economic power rarely exists; the structure and organization of firms within capitalist economies is such that they can influence the behaviour of the other elements. In a spatial context firms, which are the main productive and decision-making agents (Chisholm 1970, 7), effectively decide the geographical structure of capitalist economies.

The functions of firms
One of the sources of the economic power of firms is the essential organizational link that they provide between consumers and resources. This function involves firms in several sets of interrelated activities. First, they must attempt to assess the needs and desires of consumers as expressed by the demand in the market for goods and services, assess their own abilities to provide for this demand, create alternative demands or extend their technical ability for production. Secondly, by using appropriate criteria of judgment they must make four sets of inter-related production decisions (fig. 5.1): (i) the range and type of goods to be produced; (ii) the quantity to be produced; (iii) the techniques to be used; and (iv) the location of production. Thirdly, firms must organize and execute the production thus decided upon. Finally, they must undertake or at least initiate the distribution of their products to the consumers. All these functions cause firms to take executive action in the form of decision making, and technical action in the implementation of the decisions.

The process of production involves firms in the combination of factors of production (the flow of productive services derived from the resources of labour, land and capital) to create goods and services demanded by consumers. These factors of production are known as inputs to the production process and they are used in combination to produce outputs. Inputs are the source of firms' costs and outputs the source of revenue. The costs of using any input may be measured by the opportunity cost of not using it in its best alternative use. Thus, if the objective of firms is to maximize profits then, in making the four sets of production decisions they must attempt to maximize the difference between revenue and opportunity costs. As opportunity costs measure the value of factors of production in alternative uses, the existence of

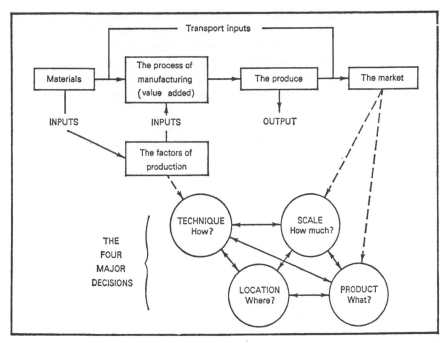

5.1 *Decisions and the process of production.* (*Source: based on Smith 1971.*)

profits in a particular industry will serve to attract factors of production towards it and away from less profitable alternatives and so will contribute to the re-allocation of scarce resources.

Decision making and the process of production

With this brief introductory background it is now possible to examine each of the four major production decisions in turn. However, it must be emphasized at the outset that each of these decisions is closely interrelated with the others (fig. 5.1). Thus choice of output may be affected by the location of a firm if the demand for its production is localized, whilst size of demand within the accessible market area may condition the scale of output which, in turn, can help to determine the techniques of production. As a result all the production decisions should be made simultaneously although for ease of exposition they will be considered sequentially in the following discussion.

CHOICE OF OUTPUT

If the objective of firms is to maximize profits then they will choose to produce outputs yielding the highest current or potential profits, a major determinant of which is consumer behaviour and the level of demand. However, the entry of a firm into an industry may increase the level of output of goods from the industry and, assuming that demand remains constant or increases less than supply, prices will fall and with them revenue and profits. This will help to discourage future entrepreneurs from establishing firms in the industry and the size of its output will stabilize. In this way the sectoral structure of an economy is controlled by the relative profitability of individual industries. Sectoral groups of profitable industries expand their use of resources and output at the expense of less profitable industries.

Choice of output is shown diagrammatically in fig. 5.2. The first of

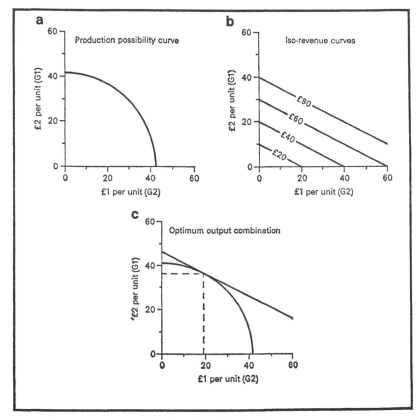

5.2 *Choice of output.*

these graphs (a) shows a production possibility curve tracing the various combinations of two products G_1 and G_2 which the firm may produce with the resources at its disposal. If the firm uses these resources efficiently, the last resources to be used in the production of each commodity will be the least suitable. As a result, the curve is concave to the origin because the first resources released from G_1 in order to expand the output of G_2 may detract little from G_1 but contribute much to G_2. The output of G_2 increases very rapidly at first as the resources most suited to its production are released, but as less suitable resources are used, it requires more and more to sustain the expansion of output and so cuts more rapidly into the output of G_1. Fig. 5.2b shows a family of iso-revenue curves which indicate the amount of revenue that would be earned by alternative bundles of G_1 and G_2. Thus 20 units of G_1 at £2 per unit would earn as much as 40 units of G_2 at £1 per unit. In fig. 5.2c the production possibility curve is superimposed upon the iso-revenue curves, the optimum level of output being determined at the point where it is tangential to the highest iso-revenue curve. In this case the optimum level of output is 38 units of G_1 and 19 units of G_2. In this way the structure of output, with a given production possibility curve, is determined by the ratio of product prices.

CHOICE OF TECHNIQUE

It is normally possible to produce a given output in several different ways and firms are faced with a choice of production method. Techniques of production vary according to the ratios in which the resource inputs are combined. For example, it is quite possible to produce a given output of an agricultural product by using a large amount of land with relatively little labour and capital (extensive farming) or by using a small amount of land combined with relatively large amounts of labour and capital (intensive farming). The choice of production technique operates under the principle of substitution which states that cheaper substitutes will replace more expensive factors.

The cheapness or expense of a factor for use in production depends upon its unit price and upon its productivity, which is a measure of the amount of the factor used in the production of a unit of output. If, over time, less of the factor is necessary for the production of a given output, its productivity is said to be increasing; whilst if it is necessary to increase inputs of the factor to maintain output, its productivity is said to be decreasing. However, factors of production are normally used in combination for productive purposes and the productivity of a single factor is in reality impossible to assess on a comparable basis without taking into account the quality and quantity of the other factors in use. Thus any comparison of the productivity of labour between firms or economies in, say, the motor-car manufacturing industry is incomplete without consideration of the influence of variation in type, quantity and quality of the machinery used by the labour forces under analysis. Variations in output per unit of labour can just as easily be caused by variation in the suitability of machinery operated by the labour force as it can by the quality of the labour itself. Because of this, it is often more appropriate, when analysing choice of technique and input combinations by firms, to consider the productivity of alternative combinations or bundles of factors rather than single factors in unreal isolation.

The way in which a firm might choose its combination of variable inputs is illustrated in fig. 5.3. Once again, diagrammatic expediency

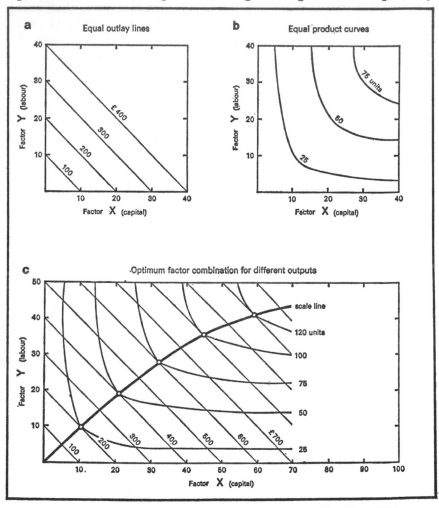

5.3 *Choice of production technique.* (*Source: Smith 1971.*)

limits the discussion to the choice of two factors only, but the method of reasoning is easily applied to three or more factors. In fig 5.3b equal-product curves show the alternative combinations of labour and capital which may be used to produce given levels of output. The equal-product curves indicate that labour and capital are substitutes but they also show that a certain minimum quantity of both labour and capital are required for production – at least 3 units of labour and 5 units of capital are necessary to produce 25 units of output. Hence the slope of the curve is convex towards the origin of the graph. A shift of the equal-product curve towards the origin represents an increase in productivity because it indicates that a smaller bundle of the input factors is needed to produce the specified level of output. A shift away from the origin indicates a decrease in productivity. In attempting to minimize costs the firm should either be directly involved in research and development or be receptive to innovations in production techniques which push the equal-product curve to the left.

The precise combination of factors used to produce a given output at any one time depends additionally upon the relative prices of the factors. If the price of capital is £20 per unit and that of labour £10 per unit, then for £100 outlay the firm may obtain 5 units of capital or 10 units of labour or a smaller quantity of both factors. The combination of factors that may be purchased with selected outlays is shown in fig. 5.3a by equal-outlay lines. Maximization of profit involves the firm in the choice of technique which, for a given level of output, has the highest productivity and which uses relatively large amounts of cheaper factors and small amounts of the more expensive factors. The optimum input combination can be derived by superimposing the equal-outlay lines upon the equal-product curves; the point at which the desired equal-product curve touches the lowest equal-outlay line is the optimum input combination.

Fig. 5.3c shows that for an output of 25 units of product the minimum cost is £200 using 11 units of capital and 10 units of labour. Any other combination of factors (e.g. 20 units of capital with 6 units of labour) would increase cost outlays without increasing production. If the relative price of labour and capital changes, then the optimum input combination will change; whilst a change in productivity of one or both of the factors will also alter the optimum input combination and change the minimum cost-outlay for a given level of output.

An increasing volume of output is often associated with the increasing use of capital plant and machinery. For example, as a farmer expands his acreage under grain it may be possible for him to substitute a larger for a smaller combine harvester and so release labour formerly necessary to attend the smaller machine. This represents a special case of substitution and in fig. 5.3c it is clear from the scale line that the optimum combination of labour and capital changes as the scale of output increases. In this case capital is substituted for labour, high capacity machines for low capacity physical labour.

One problem which fig. 5.3 does not show is that capital investments are often indivisible or 'lumpy'. This means that, although it would be most convenient to substitute capital for labour in small increments along the scale line of optimum input bundles, in the short run (see below) at least, the firm must usually decide whether to intensify its use of variable resources or to under-use a high capacity piece of capital equipment. For the small firm (e.g. a farm) this problem can be severe and this is one reason why the agricultural industry has developed institutions like production cooperatives which enable groups of farms to share large pieces of capital equipment and buildings too large for them to use economically as individuals. The provision of spatial infrastructure also involves the problem of large minimum quantities of capital and we shall come across this problem again in the discussion of transport networks (chapter 9).

CHOICE OF SCALE OF OUTPUT

Having made decisions about choice of output and production techniques, the firm is faced with the choice of scale of output. How much of its chosen products should it produce? For the profit maximizing firm the answer is that it should produce at that level of output which maximizes the difference between revenue and costs. As output increases profits will continue to rise as long as the additional cost of producing the extra units of output (the marginal cost) is less than the extra revenue received for them (the marginal revenue). As soon as the marginal costs exceed marginal revenue then total profits will begin to fall. From this it follows that maximum profits will be made at the scale of output where marginal cost is equal to marginal revenue.

Costs, generated by the firm's expenditure on inputs, may be expressed as total costs and average costs as well as marginal costs. Total costs are simply the firm's absolute total of expenditure on inputs whilst average costs are total costs divided by output, thus expressing the level of cost per unit of output. In the short run a firm is unable to vary all its inputs and so can change its level of output only by changing the level of variable factors. Under these circumstances a limitation is placed upon the output flexibility of the firm as the principle of diminishing returns operates. This principle states that, as certain factors of production are increased whilst holding others constant, output increases rapidly at first but after some point becomes smaller and smaller and may even start to decrease as too many of the variable factors are employed. It follows from this that at some level of output the cost per unit of product must be at a minimum and that short-run deviations from this level of output would increase unit costs. Thus the short-run average cost curve, which plots average cost against level of output, is characteristically 'U' shaped. The long run is defined as the period in which all of the firm's inputs are variable. It is not a fixed length of time and varies with the technology employed by firms. The generation of electricity is a well-known example of a supply industry which needs to plan increases in capacity some five or six years ahead of demand because of the lengthy processes of political, technical and locational decision making and plant installation which precede the commissioning of a power station. Associated with the short run and long run are the concepts of fixed and variable costs. As costs derive from a firm's inputs it follows that all costs are variable in the long run; only in the short run can some costs be regarded as fixed and they are sometimes known as overheads.

The long-run relationship between levels of cost and levels of output may be illustrated by long-run cost curves. Four examples of long-run average cost curves are shown in fig. 5.4. If average costs decrease with an increase in output the firm in question is said to exhibit economies of scale. By contrast, a firm whose average costs increase with an expansion of output is said to suffer from diseconomies of scale, whilst a firm whose

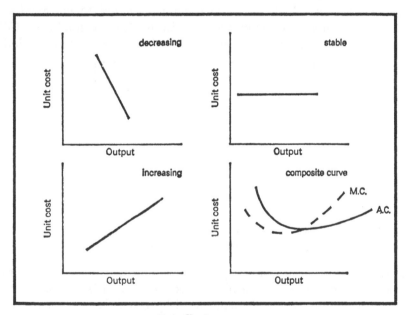

5.4 *Cost curves.*

costs remain constant with changes in output is said to exhibit constant returns to scale. Normally the long-run average cost curve of any firm is compounded of all three types of curve. At lower levels of output the firm faces economies of scale; at intermediate levels constant economies; and at higher levels diseconomies of scale. This sequence results in the characteristic 'U' shaped long-run average cost curve. Fig. 5.4 also illustrates the similar shape of the marginal cost curve (MC). As long as the marginal cost of producing the extra output (marginal output) is less than average cost for the previous level of output, then average cost (AC) is bound to fall; but as soon as the marginal cost is greater than average cost then it is similarly bound to rise.

Increases in output also affect the revenue of the firm. If the level of output is small and the utility of the product is high, then consumers will be willing to pay a high price per unit of product. However, as output from the firm increases and the marginal utility of consuming extra units of product diminishes, demand will fall and so prices and the firm's revenue will also fall. The relationship between revenue and output is shown in fig. 5.5 which also distinguishes between marginal and average revenue. Marginal revenue (MR) is the extra revenue derived from selling one extra unit of product (the marginal unit) whilst average revenue (AR) is total revenue divided by total output. As demand decreases marginal revenue falls and so drags down the level of average revenue at the same time.

By superimposing the revenue and cost curves it is possible to find the level of output at which marginal revenue is equal to marginal cost and hence at which profit is maximized. In fig. 5.6 marginal costs (MC) rise to equal marginal revenue at an output of over 7 units so that 7

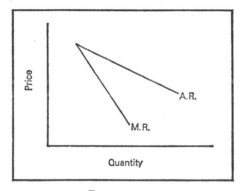

5.5 *Revenue curves.*

(the nearest whole unit quantity) is the optimum output. Average profit (£9) is the difference between average revenue (£18) and average cost (£9). Total profit may be calculated by multiplying average profit by the level of output. In this case total profit is £63 and this is its maximum level because any increase or reduction in output will reduce this figure.

Normal profits are calculated inclusive of all costs, including payments to shareholders to cover the opportunity cost of their capital and so prevent it from being transferred to other firms. But it is clear from fig. 5.6 that profits in excess of all costs are being made. These are known as super-normal profits and in a perfectly competitive situation would be a signal for firms to move resources into the industry in which the excess profits were being made. Such a situation is shown in fig. 5.7A.

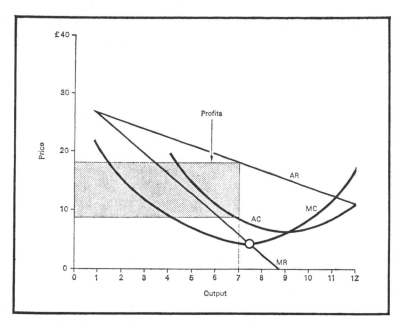

5.6 Output determination under monopolistic conditions.

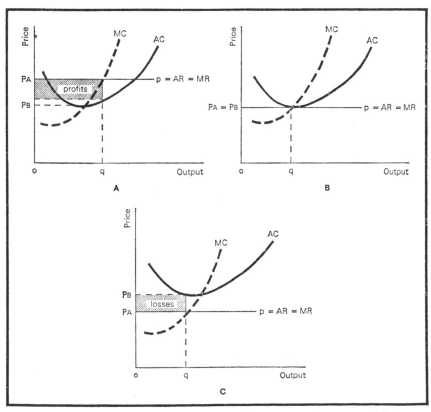

5.7 Output determination under perfectly competitive conditions.

Here it will be seen that price (p) is equal to average revenue (AR) and marginal revenue (MR). This is because the output of each firm is so small relative to the total output from the industry as a whole that a change in the output of any one firm could not affect supply in any noticeable manner. It follows that no individual firm is able to affect the price level by its own actions. Firms, in perfect competition, are economically powerless with respect to the price level; they are price takers.

In fig. 5.7A super-normal profits are being earned by a firm equating marginal costs and revenues whilst in fig. 5.7C losses are being made (average costs are greater than average revenues). Under conditions of perfect competition, characterized by the free entry and exit of firms, the situation in (A) would attract firms to the industry whilst (C) would repel them. The collective effect of a large entrance to or exodus from an industry would affect the price level by substantially affecting the supply; the long-run equilibrium of the industry and hence of each individual firm is shown in fig. 5.7B where all excess profits have been eliminated and neither entry nor exit is stimulated. The structure of the agricultural sector approaches conditions of perfect competition with a large but diminishing number of sellers and buyers. However, the many partially uncontrollable fluctuations in agricultural price levels result in substantial government interference in the market for agricultural products and so modify the influence of perfect competition upon price formation.

More generally in a spatial context, the existence of distance and transport costs restricts freedom of action on the part of both buyers and sellers (see pp. 37–44) and so automatically removes one of the assumptions of perfect competition and replaces it by monopolistic or imperfect competition. This may be defined as a market situation in which there is a large number of firms whose outputs are similar but not perfect substitutes because of deliberate product differentiation or differentiation induced by locationally separate market areas. It is distinguished from perfect competition by the fact that there are certain, albeit small, barriers to entry so that the output decisions of individual firms significantly affect total supply and the average and marginal revenue curves slope down from left to right as output increases (fig. 5.6). The assumptions of monopolistic competition are very important for economic locational analysis because space restricts entry and exit, markets are spatially fragmented, and any firm has a degree of monopolistic control over certain spatial sectors of the market. However, any excess profit in a monopolistic situation may attract competitors able to overcome the barriers by locational adjustment or persuasive advertising and they may whittle away these profits by shifting the AR and MR curves to the left. But an AR curve sloping down from left to right can only be tangential to an AC curve at a point below

optimum output and above minimum cost. This means that monopolistic firms produce a quantity of output below that of their most
efficient level of production at the point of minimum average costs and
so incur waste even when super-normal profits are removed by competition.

CHOICE OF LOCATION
For the economic geographer the theory of the firm is deficient because
it does not normally consider spatial variations in marginal revenue and
cost curves. This is an important omission as both demand (measured
in terms of economic population potential) and the availability of
inputs (in terms of cost) vary substantially from place to place (Smith
1971, especially chapters 3 and 4). As a result of this spatial variation,
profit-maximizing firms must attempt to choose an optimum location as
well as an optimum output and technique of production.

Location decisions may be regarded as problems in (i) spatial relations
or (ii) investment allocation. (i) The least-cost approach, enunciated by

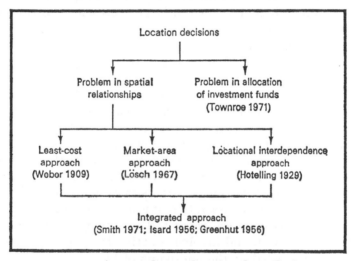

5.8 *Approaches to locational analysis.*

Weber (1909), took account of the uneven distribution of factors of production in space but made the limiting assumption of the existence of a
perfectly competitive market in which price is unalterable by individual
firms. Under these circumstances, Weber pointed out that the most
profitable location would be at the least-cost position, which could be
determined by calculating the costs of transporting inputs and outputs
to and from the plant. Such a transport-oriented least-cost location
could be modified if savings could be made by locating within reach of
a source of cheap labour or at a location where external economies of

scale compensate for the increase in transport costs generated by the deviations from the transport-oriented optimum location.

The most important limitation of Weber's approach is the assumption of perfect competition. As has been noted earlier, the friction of space itself prevents perfect spatial competition, as a firm can command a nearby market area in which proximity allows it to undercut its competitors. Weber overcame this problem by regarding the market as a point rather than as a continuous distribution. But the demand effects of a spatially dispersed population in generating a market area were demonstrated by Lösch (1967) and his spatial demand cone has already been mentioned (p. 36). Nevertheless, it should be pointed out that although Lösch used the idea of imperfect competition in space he did not allow for an uneven distribution of factors of production because he assumed the existence of an isomorphic plain. In this sense he complemented Weber's approach although his assumption is a limitation upon the empirical usefulness of the theory. This is especially true of an era in which transport costs are diminishing and so increasing the size and degree of market area overlap and in which output is increasingly concentrated in a limited number of large firms selling to national and international markets.

Although the locational interdependence approach also ignores cost factors, it does take into account the impact of competitors upon firms' location decisions. Hotelling (1929) analysed the relationship between two firms serving a linear market – two ice-cream sellers on a beach! He assumed conditions of inelastic demand, spatially uniform production costs, no costs of relocation, but distance-induced transport costs. It was suggested that price competition between the firms would be unworkable, so that the only alternative was locational adjustment. A series of unstable locational situations would eventually result in the two sellers standing back to back, each serving one half of the market. Whilst this is the most stable solution, it is not the most efficient, because excessive transport costs are generated in serving distant consumers. An optimal solution would involve splitting the linear market in half with each firm located at the mid-point of his market area. But this would be unstable, there being the temptation for either firm to move towards his competitor to compete for the patronage of his competitor's most distant customers.

Each of these approaches is frustrating in the sense that they incorporate many limiting assumptions, and these assumptions tend to complement each other. Smith (1971) provides an extensive review of the major industrial location theories, including the attempts to construct an integrated theory (Isard 1956; Greenhut 1956). In addition he presents an original synthesis, the great value of which is that it is based on a series of simple concepts and attempts to stress the relevance of industrial location theory for solving real-world problems, as opposed to abstract theoretical problems.

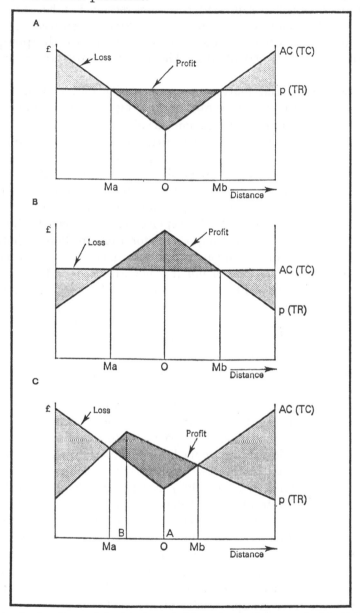

5.9 *A simple model of industrial location.* (*Source: Smith 1966.*)

The essence of Smith's approach is shown in fig. 5.9. This embodies the assumptions that firms are price-takers, their output is constant in space, and that spatial variations in the level of demand are reflected by spatial variations in price. Fig. 5.9. represents a series of cross sections through two-dimensional surfaces representing spatial variations in cost and price. They use the measures of average cost (AC) and average revenue or price (p) to represent the cross section but because of the assumption of constant output and revenue each measure can be transformed into total cost (TC) and total revenue (TR) with appropriate adjustments of the vertical axis. The diagrams are largely self-explanatory. They show an optimum location (O) at which profits are maximized and spatial margins to profitability (Ma and Mb) where average revenue is equal to average cost. In fig. 5.9c maximum profit is earned at the minimum cost location (A) although an alternative configuration of the cost and revenue surfaces may shift the optimum location to the point of maximum revenue (B).

The concept of the margin to profitability was first introduced by Rawstron (1958). It suggests that firms can make profits and stay in business without finding the optimum location – a site normally quite restricted and hence difficult (expensive of resources) to find. In a dynamic situation it is almost impossible for a firm to know whether it is at the optimum, but Rawstron argues that its profit and loss account would soon indicate whether or not it was located beyond the margin. Thus the spatial margin introduces an element of realism into location theory and in so doing it loosens up the spatial environment within which firms operate. In short, the concept of the margin creates a dynamic space within which economically sub-optimal locational decision making by firms can be accommodated.

The most explicit spatial example of sub-optimal decision making by firms is Wolpert's (1964) oft-quoted comparison of optimum and actual patterns of labour productivity on farms in middle Sweden. On average, the farms in the area operated at about two-thirds of their potential productivity although the spatial configurations of actual and potential productivities were remarkably similar. Such sub-optimal behaviour by firms is the result of imperfect knowledge and a limited ability to use information (Pred 1967). Even if knowledge is freely and abundantly available to firms the behavioural or perceptional environment of individuals or groups of decision makers filters out much of the data transmitted to them from the objective environment and so is responsive only to certain types of recognizable data (Kirk 1951, 1963). Furthermore, distance and the hierarchical structure of spatial organization, combined with the effect of well-defined social groups which limit the inter-group transfer of knowledge, together constrain the flow of information through space. Thus, in H. A. Simon's (1952) terms, decision makers in the real world operate under conditions of bounded rationality rather than the global rationality assumed by economic man.

Once received, knowledge must be used by a firm. Large concerns may exploit internal economies of scale from the provision of specialized data-gathering and processing units (e.g. Steed 1971a); but in small firms, typified by the farm-firm (Munton 1969), this function must be added to the more immediately important day-to-day functions performed by a limited number of managers. Time must then be found to make decisions on the basis of the processed information, and here again the design of management structures affects the quality of the outcome although feedback from previous situations can be productive if it is used as part of a learning process which can help to standardize decision-making mechanisms. In addition the personal characteristics of the decision makers and, in particular, their attitudes to risk under conditions of uncertainty closely affect the type of decision made. Simon (1952) suggests that decision makers are satisficers in that they rarely bother to search for, or to define, the optimal solution but merely accept satisfactory alternatives and reject the unsatisfactory. This concept is given credence by the example quoted earlier (Wolpert 1964). The greatest difference (up to 90%) between the pattern of actual and potential labour productivity on the farms in middle Sweden was found in the zones of high actual and potential productivity. The relative relief of the actual surface was more subdued than that of the potential, which suggests that a satisfactory rather than an optimum return was the goal of farmers in this area.

The great scope for sub-optimal decision making underlines the importance of Rawstron's (1958) marginal concept for studying the evolution of the locational structure of firms. The fact that margins may be as difficult to define as are optimum locations (Chisholm 1970) is much less important than the substantial analysis of industrial location (Smith 1971) which may be built upon the concept.

(ii) A different approach to industrial location decisions is adopted by Townroe (1971). He points out that location decisions are in fact investment decisions and are only one of the four major sets of decisions faced by firms, each of which involves the allocation of investment funds. In this context, the critical variables in the location decision become the overall policy and investment strategy of the firm (e.g. Watts 1971) and the time period available for making the decisions. These variables provide the criteria for evaluating the pressures necessitating a decision; this evaluation process is important because any of the four major sets of decisions can act as imperfect substitutes for each other in adapting to changed conditions. This may be exemplified by Krumme's (1969) discussion of the constraints of time upon decision making. In the short run locational change might be impossible and another form of adjustment – for example a change in input combinations or product-mix – may be necessary. In the medium term, whilst relocation is still unlikely, expansion and contraction at existing plants may be adopted as a solution and so indirectly affect locational patterns, but in the long run full locational adjustment is much more feasible. This approach to locational decision making is more broadly based than the space-relationships approach. It recognizes explicitly the interdependence of the major sets of decisions, and it lays stress upon the variety of constraints upon locational and other investment decisions, exercised by factors internal to the firm, rather than concentrating solely upon the role of space.

The challenge of large firms

The emphasis in economic geography upon the reactions by firms to a variable spatial environment and their subsequent organization of space has been couched mainly in terms of perfect and monopolistic market conditions. Economists recognize two other market situations, monopoly and oligopoly, the implications of which have, with a few exceptions (e.g. McNee 1958, 1960), been ignored by geographers. Monopolies exist when one firm constitutes an entire industry by selling a product for which there are no close substitutes. They are often protected by entry and exit barriers such as a large minimum size of operation which results in important and beneficial economies of scale but requires vast quantities of capital to begin production. Clearly, total supply is directly affected by a monopoly firm's activities so that the process of output level and price formation is similar to that shown in fig. 5.6, with the important proviso that the barriers to entry prevent the elimination of excess profits because competitors are not free to enter the market in order to increase total supply and so shift the AR and MR curves to the left.

An oligopoly exists when a few firms dominate an industry. The major characteristic of such a market situation is that, like monopoly, the output decisions of each firm directly affect total supply, but in addition the demand for one firm's products is closely affected by the actions of the others in the market. This situation is normally associated with industries characterized by large firms (e.g. the petroleum industry), but this is not necessarily so. The discussion of oligopolistic price collusion by retail outlets in a locationally isolated market area (pp. 40–41) revealed how space can act to induce oligopolistic conditions. Its essential feature is the mutual interdependence of all the major producers in an industry which, because of the small number of firms involved, allows and stimulates them to act collectively as a monopoly.

Within a national economy, monopoly and oligopolistic firms clearly have great economic power in that they effectively control their market for goods and services – a fact recognized in many capitalist countries by legislation designed to control abuse of that market power. However, the ability of large firms to circumvent this attempt to control their activities is well exemplified by the early development of the Gröningen natural gas fields in the north-eastern Netherlands (Odell 1969). The importance of this gas field derives from its size and its location in north-west Europe adjacent to the world's geographically most concentrated zones of energy use. Given many producers as well as many consumers it is probable that the output of the field would have been competitively bid up to its technically optimum level and the price reduced until a customer could be found for the marginal unit of profitable production. But the contrast between possible and planned production levels and a pricing structure enabling super-normal profits to be earned leads to the conclusion that output levels and price were determined in ways other than by the competitive forces of a free market. The massive size of the gas field and the wide range of production possibilities imply that its marginal cost curve could be considered to be horizontal. In a monopoly situation the firm's problem then becomes that of choosing the level of output which will maximize profits. This is an exercise in applied economic geography because the marginal revenue derived from sales to consumers in different locations needs to be converted into f.o.b. prices. But, as the costs of pipeline transport depend upon the distance and volume of gas moved, this exercise requires a substantial knowledge of the Western European energy market. This latter consideration underlines the economic power of large vertically integrated firms (Shell and Esso in this case) in that their knowledge and understanding of their industry is without equal. The output level which maximized profits and protected the firms' previous European investments in oil refining and distribution was considerably lower (an estimated 50% lower) than the possible potential production. It involved the distribution of gas to premium consumers and not to the low price bulk energy market in West Germany, or the iron and steel industries of the Saar, Luxembourg and eastern France.

The internal policies of large firms are also powerful influences upon the allocation of resources. The increasing dominance of large multi-plant, multi-enterprise firms in the British economy (Parsons 1972) has the effect of imposing an internally controlled and semi-autonomous network of production and decision-making facilities over the regions of Britain. This means that the local benefit deriving from the regional location of part of a large firm may be limited, as many of its functions are undertaken not in the region but at a central specialized unit. But the spatial distribution of these decision-making units of large firms is far more concentrated than is the distribution of their operational units and shows a distinct preference for the South-East and West Midlands regions. This trend is likely to increase as national rather than regional firms take control of more sectors of industry (Watts 1972). More generally, Steed (1971b) has pointed out that the location, growth and trading policies of large firms have fundamental implications for the

D

spatial structure of economies, economic growth and the bases of international trade.

The fact that the world economic space is partitioned into politically defined national economies with a resultant mosaic of economic controls and rules of behaviour, provides multi-national firms, operating in a number of countries, with a further source of power (e.g. Steuer, *et al.* 1973). Such firms may take advantage of differing tax systems by declaring the majority of their profits in countries with low rates of taxation. Or, like the major international petroleum companies, they may exploit their vertical integration to prevent the independent marketing of petroleum by the countries in which the major reserves are located. Through their location strategy, they may also play one country or group of resource owners off against another in the attempt to secure concessions. Whilst most economies suffer both from this type of manipulation and from the reduction in national political control over foreign owned investments, it is inevitable that the less-developed economies suffer most. This is because of their lack of bargaining power and their economic inability to cope with uncompetitive pricing, the import of processed commodities, the export of unprocessed raw materials and the remission abroad of a large proportion of profits resulting from the operation of foreign owned companies.

As the power of firms transcends national boundaries it is clear that international economic cooperation is a necessary form of counter-vailing power. However, the differing aspirations and vested interests of individual parental and host nations makes such cooperation difficult (Penrose 1968), as the effective but troublesome operation of the Organization for Petroleum Exporting Countries (O.P.E.C.) shows. This is the best-known example of international cooperation and is designed to enhance the value of the oil reserves located within its member states. Success has been achieved by adopting unified policies towards the major oil companies, thereby increasing the bargaining power of the individual states, not only to secure higher prices and royalties but also to gain an increasing degree of control over the companies' operations in the exploitation of the oil reserves. Furthermore, the Arab countries of the Middle East can use their new found economic power for political ends by denying oil to Israeli sympathizers for as long as support for Israel is maintained. But, because of the vertically integrated nature of petroleum firms, the producer countries may find it necessary to impose a general embargo upon the export of oil, or engage in inter-governmental bargaining, to prevent its passage to any particular market via the major oil companies. In the short term, however, O.P.E.C.'s policy has been very successful. Between 1970 and 1974 the value of the organization's exports rose by 663 per cent and the value of its total trade by 512 per cent.

Conclusions

This chapter has introduced the major types and processes of decision making by firms. Its sequential presentation should not obscure the fact that the major decisions are interrelated, and the positive bias which has revealed the main processes of decision making should not be allowed to imply that the decentralized character of decision making in mixed or capitalist economies is necessarily the most effective or the most efficient. Firms are not only decision makers, they are also increasingly important sources of economic power. This power derives, in part, from the ability of firms to derive monopolistic advantage in space, but the increasing size of firms and their increasing share of the market for many products are the main sources of this power.

If, as Galbraith (1967) and Chisholm (1970) argue, large firms are the basic planning units in capitalist economies, playing the major role in the allocation of scarce resources into alternative uses, then this shift of economic power towards large firms has serious implications for the nature of society. Whilst many benefits flow from size, especially those associated with economies of scale and the ability to finance expensive technological research, consumer choice is restricted and manipulated and resource employment becomes increasingly dependent upon control remote from and independent of local environments which form the milieu for the lives of resource owners.

6 Resources

The purpose of this chapter is to show that the operation of an economy both creates and destroys resources and so has fundamental implications for the distribution of income to resource owners. Resources are inputs to the process of production and they generate a flow of productive services, known as factors of production, when used in combination by firms. The distribution of income amongst resource owners in a capitalist economy is a function of the concentration of resource ownership and the evaluation of resources by firms.

Convenient groupings of resources
The resources of an economy are frequently classified into three groups: (i) *land*, including geographic space as well as natural resources; (ii) *labour*, or the manual and mental abilities of the economically active population; and (iii) *capital*, comprising both fixed capital, those artifacts deliberately made by man for use in the production process, and investment capital which provides firms with the ability to extend their claim over fixed capital. Technology, defined here as the knowledge of and ability to implement alternative techniques of production, is often considered as a separate resource but the supply of technology may derive from individual inspiration (labour resource) or from a planned process (research and development) designed to increase knowledge. In this latter case technology is more correctly thought of as a capital resource in that it is produced for future use by diverting resources away from the immediate production of consumer goods. In fact, each of the resources used by firms is a combination of the three groups: labour is educated and trained, land is cleared and made more accessible, and capital is manufactured by a combination of resources. Whilst it is pedagogically convenient to group resources into typological sets the boundaries of the sets intersect and are very blurred.

Demand and supply
As inputs to the process of production, resources can be of economic service only if there is an effective consumer demand for the goods and services they are employed to produce. But resource creation by such derived demand is effective only if it is sufficiently powerful to provide

profit maximizing firms with an incentive to incur the costs of resource use or if, by adopting other criteria, governments deem the employment of resources to be worthwhile. Thus, although knowledge about the existence of resources is an essential prerequisite of their employment it is not of itself sufficient. Known stocks of resources have no inherent worth of their own; they are created in the economic sense by the interaction of their derived demand and supply conditions, themselves a complex function of ecologic, ethnologic and economic forces (Firey 1960).

DEMAND

Demand for resources emanates from the four production decisions of firms discussed in the previous chapter. The total demand for resources is a function of the volume of output produced by firms in the economy and of the efficiency and technology with which resource inputs are combined to produce that output. During periods of depression the aggregate demand for goods, defined as consumer plus investment demand, falls and the production of goods and services declines, dragging down with it the level of resource-employment and income. The reverse is true in conditions of prosperity when aggregate demand and output increase and resource demand is pulled up. Such fluctuations in the level of economic activity are known collectively as the business cycle and they are a feature of capitalist economies which rely upon the interlocking confidence between firms to make profits and to maintain or increase the level of flows through the economy. Indeed, the business cycle is one of the major influences upon resource demand and the most effective cause of resource unemployment. Ever since the intense international resource unemployment associated with the Great Depression, which began on the New York stock exchange in October 1929, capitalist governments have attempted to mitigate cyclical fluctuations in the economy; their attempts to prevent high levels of such unemployment have proved to be one of the most important components of government economic policy in the capitalist world.

The integrated nature of economies is a major factor in the generation and spread of depressions. A fall in aggregate demand results in a fall in the demand for resources which reduces income and so reinforces the initial decline in demand by further reducing consumer demand. The sectoral spread of a depression through an economy may be predicted by a series of leading indicators which regularly precede (lead) or follow (lag) a national business cycle and this concept has been extended to the study of leads and lags over the regions comprising a space economy. The potential of this integration, between measures of temporal process and spatial structure, for spatial economic planning and forecasting is considerable: it could identify the spatial leaders of local business cycles, provide details of the spatial transmission of a business cycle through a system of regions, reveal regional variations in the susceptibility of local economies to the business cycle and measure the spatially variable effects of national as opposed to regional influences upon the generation of business cycles. At a national scale Brechling (1967) has shown that the Midlands, London and South-East regions of the United Kingdom are leaders of the business cycle, whilst Scotland, the North and North-West regions lag. In the United States King, Cassetti and Jeffrey (1972) show that the mid-western cities consistently lead the nation in generating fluctuations and are also particularly susceptible to national rather than local cyclical factors. At a more local level Sant (1973) found areal similarities to be more effective than hierarchical links in controlling the spatial spread of business cycles in East Anglia. But in south-west England spatial influences are masked by seasonal, holiday-induced, employment fluctuations (Bassett and Haggett 1971; Haggett 1971).

However cycles generated by inter-city interactions, rather than by national influences like major economic policy decisions, are likely to be spatially selective in their impact upon cities because of the differential hierarchical relationships, economic interdependence and distances between them. Thus in the mid-west of the U.S.A., three clusters of cities based upon Pittsburgh–Youngstown, Indianapolis and Detroit have been identified on the basis of strong and simultaneous effects of cyclical fluctuations in employment induced by high levels of interaction, whereas certain other cities in the area – being less well integrated – tend to remain remote from such fluctuations (King, Cassetti and Jeffrey 1969).

The local effect of nationally-generated cyclical fluctuations may also be influenced by the composition, size and diversity of economic activities located within each region. However, recent evidence suggests that in the United Kingdom at least, regional differences in the cyclical sensitivity to employment over the years 1949–64 were due not so much to inter-industry (compositional) factors as to intra-industry factors. These relate to the notion that fluctuations in the unemployment rate of given industries are not the same in all regions (Harris and Thirlwall 1968). These authors conclude that detailed analyses of interregional variations in the performance of individual industries would seem to be more relevant for the study of regionally variable fluctuations in employment rates than are studies of the industrial composition of regions.

Long-term changes in the structure of demand for particular products have a more sectorally and locationally restricted effect on the demand for resources than have changes in demand resulting from the business cycle. The decline in the demand for coal in Western Europe with the increasing attraction of petroleum during and after the late nineteen-fifties is a classic example of change in resource appraisal due to product substitution. Labour, land and capital resources within the coal industry have been devalued, and the resultant structural unemployment of these resources is often irreversible as coal mines cannot easily be brought back into production once they have been abandoned.

The substitution of one resource for another in the interests of production efficiency is another important influence upon the demand for resources. The tendency for capital to replace labour as the size of firms increases is a well-known example of this process. Expansion of the capital input to the production process has resulted from the technological creation of capital resources embodied in mechanized and automated means of production. Substitution of this sort, limited as we have already seen by the principle of diminishing returns (p. 70), involves the re-evaluation of labour resources because fewer units of labour are required to produce a given level of output. But it can also diminish the qualitative input of labour by reducing the element of skill in jobs and by limiting labour participation to one small step in the production process.

SUPPLY

Current interest in the relationship between economies and ecosystems incorporates a concern for the long-term supply of currently vital but non-renewable natural resources. This is justified to the extent that natural resources are not inexhaustible but neither are they rigidly limited. Their supply has tended to increase as the effort expended to search for them has intensified and as the technology to utilize hitherto unknown resources improves to the point at which they can be used as substitutes for existing resources or as extensions to the productive potential of the economy. Again the critical variable is resource demand, because as long as demand remains unsatisfied it is worthwhile to devote existing resources to the search to extend the resource base. The centuries of land reclamation in the Netherlands have been prompted not only by a concern for safety but also by the highly effective demand for geographical space in an affluent, densely populated country with a high birth rate, today sitting astride a zone of the highest economic potential within the European Economic Community. Current and future demands for land make increases in supply worthwhile even at the high costs of extensive, technically difficult and long-term reclamation projects.

But unsatisfied demand, reflected in high prices and the potential for resource owners to earn a high income, does not necessarily increase the supply of resources. This may be illustrated by the example of labour. Any increase in price that may follow an increase in the demand for labour may induce an increase in the size of the work force by providing the owners of labour with the opportunity to increase their income by maintaining or increasing current levels of supply. However, an increase in earning power presents the owner of labour with a choice of working more hours to increase his income, or maintaining his income level by reducing the number of hours at work and increasing his leisure time. The outcome will depend upon the owner's evaluation of the opportunity cost of time spent at work.

Owners of capital and land resources may be in the enviable position of not having to make this choice when faced with changed demand conditions. They can merely allocate their resources to firms without necessarily committing themselves to any work. Payment for the use of land is made in the form of rent and is a function of the value of the productive output gained from the land for which the owners, as opposed to the users, may be at least partly responsible according to the terms of the tenure agreement. Capital earns profits for its productive capacity and for the risk that its owners take in investing it. This is a real risk in the sense that owners of capital are usually paid infrequently and after the goods produced by using the resource have been sold, but is not any more real than the risk of unemployment faced by the owners of labour resources (usually without any land or capital on which to fall back) if profits fail or techniques of production change. The supply of capital may be altered by investment from accumulated wealth held by individuals. Savings tend to increase as income levels rise because both basic needs and luxury wants are satisfied without the use of all income. It is hardly surprising then, that high-income resource owners are a major source of capital investment from savings although institutional investors, like insurance companies, banks and unit trusts, also increase the flow of investment capital by using funds collected from a large number of individuals.

MOBILITY

We suggested above that owners may decide to withhold their resources from firms despite the possibility of increasing income by increasing the supply. This is particularly true in the case of labour resources because non-pecuniary returns complicate simple monetary gains. But similar considerations also apply to capital and land, when factors such as the degree of risk, the nature of the tenurial conditions, or the expectation of an even higher pay-off in the future may cause their owners to withhold resources from the market. The same sorts of influence also affect the owner's choice of alternative uses to which his resources may be allocated. In a simple and unreal case we may assume that resource owners are income maximizers (cf. the profit-maximizing aims of firms) in which case they will allocate their resources to those firms offering them the highest price. More reasonably it may be expected that non-pecuniary benefits are also taken into account but that, as long as these change more slowly than pecuniary benefits, owners will increase their allocation of resources to those firms offering higher rather than lower returns.

The major limiting factor acting upon this allocation process from the supply side is the immobility of resources. The concept of resource mobility relates to the ease with which factors may move between alternative uses. It has two components: occupational and geographical mobility. Occupational mobility refers to the ability to change occupations, geographical mobility to the ability to change place of work, but the distinction between the two is not clear-cut in the real world where the former often subsumes the latter. However, land is geographically immobile and occupationally mobile. But changes in land use are less easy within built-up urban or industrial areas, supporting high densities of fixed capital, than in agricultural areas where land may be more easily allocated to alternative land uses, provided that its owners and controllers are willing (Denman and Prodano 1972). Fixed capital equipment is immobile in both occupational and geographical senses, although it is usually assumed that investment capital is completely mobile, responding quickly to investment opportunities and being inhibited only by occupational or geographical differences in risk. This assumption is not justified as there are strong grounds for believing that capital immobility is generated by factors associated with the finance institutions, through which much investment capital is made available to firms, which may be less willing to invest in distant, small scale or novel projects (Estall 1972). Governments may also restrict the free flow of capital as part of a regional economic policy whilst past fixed investments by firms may add to the immobility of their investment capital. Labour is potentially mobile between jobs and places, especially over long periods. In the short term, occupational mobility is impeded by the barriers of skills, qualifications and training whilst movement between locations is impeded because of social and family ties in an area. The continual entry to and exit from the labour force, which has a turnover of about 4% per year, provides some long-term flexibility for changing its geographical and occupational structure even if all its current members remain immobile.

Factor mobility is a form of elasticity. The greater the forces increasing immobility the more inelastic is the response of factor owners to changes in price. This elasticity or degree of mobility lies at the heart of problems associated with adjustments to the continually changing conditions of resource demands in a space economy; the two extremes of geographical mobility provide the limiting cases. If factors are perfectly mobile then the locational changes in resource use induced by changes in demand and production technology would take place automatically – the space-economy would be instantaneously and continually in equilibrium. If, on the other hand, resources were completely immobile within an area then changed conditions of demand and technology would have to be met by the internal adjustment of industrial structure. In the real world a finite degree of mobility exists but is not large by comparison with the rate of change of demand and technology. As we have already seen (p. 45) there are large-scale public and private costs and benefits associated with both mobility and immobility, but the spaceless economy of perfect competition does not recognize these. Economic decision making about resource allocation in the space economy is, however, forced to deal 'not with quicksilver, but with treacle' (Brown 1972, 3).

The evaluation of resources

A distinction may be drawn between the concepts of resources and reserves (Manners 1969). When making this distinction the term reserves is limited to those resources that have current economic value in terms of potentially profitable use in the production of goods and services. With a given level of demand only those resources whose marginal costs of use are less than the marginal revenue received from the goods that they may be used to produce can be classified as reserves. The remainder are marginal or potential reserves.

Consider the evaluation of the geographical space surrounding an urban centre. If we make the simplifying assumption of an isomorphic plain upon which all points have equal linear costs of access to a central town, then the only factor which differentiates one part of the space from another is the distance between it and the town. Unless the town is completely self-sufficient and able to satisfy all its demands internally, the consumer households located there will exert some demand pressure upon the surrounding space. Demands for food, residential or recreational space emanate directly from consumers, whilst demands for natural resources and space for industrial expansion come indirectly from consumers via the firms within the town. The relative strengths of these demands would be reflected in the price that consumers or firms are prepared to pay for the use of land resources, but this price would not be the same for all locations. For residential, agricultural or industrial land uses a site close to the town might be more valuable than one further away, whilst for recreational purposes the reverse may be true. Thus, for any particular location, there may well be several competing uses, each able to earn a different return from the employment of the land because of the effect of distance from the urban centre and the market conditions within the town. Furthermore, the situation may be complicated if the land surrounding the town is owned and controlled by a set of landowners, each competing to sell his land for uses that will be able to afford to pay a high rent. How is geographical space within a capitalist economy evaluated under these circumstances?

To begin with let us assume that the only demand for the use of the land is that for the production of a single food crop required by consumers within the town. The size and location[1] of this demand is critical in determining the economic value of the land resources. This should become clear from the following formula (Dunn 1954, 7) expressing rent per unit of land in terms of a number of independent variables:

$$R = E(p - a) - Efk$$

where R = rent per unit of land
  k = distance from town
  E = yield per unit of land
  p = market price per unit of commodity
  a = production cost per unit of commodity
  f = transport rate per unit of commodity and distance

The assumptions in the model are that yield, unit price, production cost and transport rate remain constant for all locations and volumes of output and that rent is expressed net of all production costs. Given perfect knowledge in the land market R may be the actual rent paid to landlords. But normally R, more accurately labelled as economic rent, differs from actual rent. The equation may be graphed (fig. 6.1) to show the linear relationship between distance and rent per unit of land given by the formula. Fig. 6.1 shows that rents are greatest nearest the town but decline with distance at a rate equal to the product of the yield and

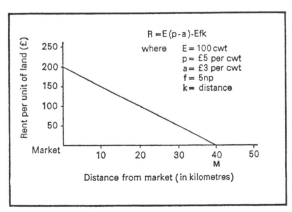

6.1   *Distance and rent. (Source: adapted from Dunn 1954.)*

[1] Presumably the population outside the town will also generate a demand for food but we will assume that this is so small, relative to urban demand, that for the purpose of this argument we can safely ignore it. A thorough discussion of the problems raised originally by the model of J. H. von Thünen is in Hall 1966.

the freight rate. The rent is entirely absorbed at point M. The sloping rent line is a function of the cost friction of distance which serves to divide the resource of geographical space into reserves, to the left of M, and potential reserves to the right of M.

The downward sloping curve is in fact a marginal rent line that tells us the extra amount that will be added to the total rent by each successive unit of land. The curve is analogous to the marginal revenue curve of monopoly or monopolistic competition. But the marginal rent in fig. 6.1 is reduced, not by the fall in product price associated with an increase in supply under monopoly, but by the cost erosion of f.o.b. prices in overcoming the distance to the market. As output is increased, by bringing more distant land into cultivation, marginal rent is reduced. As we have assumed that costs are constant with both scale of output and location and have expressed rent net of production costs, the

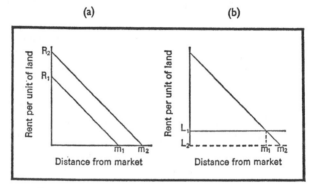

6.2 *Distance and rent under dynamic market and production conditions.*

marginal costs of production may be taken as the base-line of fig. 6.1. Beyond M marginal returns are negative whilst to the left of M marginal revenue is greater than marginal cost. Thus the distinction between reserves and potential reserves or resources is made at the level of resource allocation where marginal costs of production are equal to marginal revenues. By allocating this amount of resources total revenue is maximized. Production beyond M is not worthwhile and land-owners in this area will not receive any income.

However, the distinction between reserves and potential reserves is dynamic. Fig. 6.2a shows the effect of an increase in market price for food. If costs remain the same, or increase less than the price of the product, price increases push up the marginal rent line from R_1 to R_2 and so make the cultivation of extra land between M_1 and M_2 economically worthwhile. More distant landowners are brought into the commercial economy as their resources increase in value. A reduction in production costs, shown in fig. 6.2b by a lowering of the horizontal base-line from L_1 to L_2, could also have the effect of increasing the area of land classed as reserves from M_1 to M_2. Similarly, a reduction in transport rates would also extend the area of valuable resources by increasing the f.o.b. price for their products, but this might result in such a large initial increase in production that the market becomes over-supplied and the price and hence marginal rents fall. This would imply a reduction in the area of cultivated land, but the final limit would depend upon the price elasticity of market demand. More generally, the flattening out of marginal rent curves by reductions in transport costs reduces the economic differentiation of geographical space and increases the relative importance of factors other than distance upon the determination of rent.

The implications of competition for land between two land uses are shown in fig. 6.3a. Land-use I can earn a higher rent close to the market but is susceptible to costs of distance whereas land-use II with a lower market price earns a relatively high rent further from the market as its transport costs do not increase so rapidly. If the owners of the land resources attempt to maximize their income they will sell their land to the highest bidder. Thus, although land-user II could operate close to the market and still make a profit, he is unlikely to be able to do so unless non-economic criteria are involved in the allocation of the land. Between M_1 and the market, land-user I must pay the owner of land a level of rent at least equal to the marginal rent line of land-use II. If he does not, then land-user II can compete for the use of the land by offering the owner an amount up to the level of the marginal rent. Because it is necessary for land-user I to pay the owner of land at least that amount to prevent the land being transferred to another user, it is known as the transfer earnings of the land. So, with more than one land user, only the rent in excess of transfer earnings is known as economic rent.

Under competitive conditions the economic rent of land between M_1 and the town would be equal to the difference between the marginal rent lines of land-use I and II, described by the triangle ABC in fig. 6.3a. Rent charged at anything less than this would result in excess profits for the uses of the land and, under conditions of perfect knowledge and free entry and exit, would attract firms into the industry and so increase the competition for land and push its rent up to the level of the economic rent. But we have already pointed out that the conditions of perfect competition do not hold and that actual rents are rarely equal to economic rents. At M_1 the transfer earnings of land-use I are equal to its economic rent and beyond M_1 exceed it. In other words, the marginal rent is exceeded by marginal opportunity cost and so land-owners will sell to firms concerned with land-use II in order to maximize their income. The outermost boundary of cultivation is reached, as before, where marginal rent is equal to marginal cost.

The same argument can be extended to the allocation of land to three or more uses (fig. 6.3b). If the price of the product from land-use II_1 increases, the marginal rent line moves vertically upwards to II_2 and so its production is increased at both the inner and outer margins

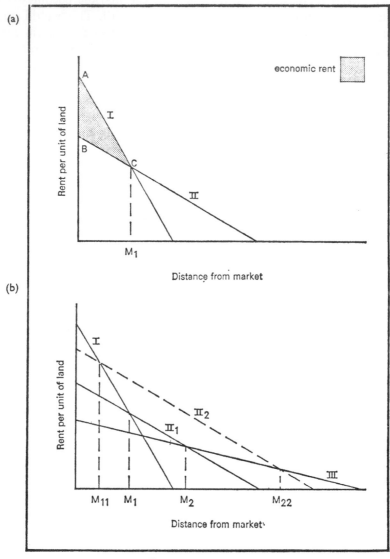

6.3 *Competition for land.*

simultaneously (from M_1 to M_{11} and from M_2 to M_{22}). This means that production from I and III is diminished and, provided demand for their products remains unchanged, the prices of products I and III will rise as consumers compete for scarce supplies. This will cause vertical shifts in their marginal rent lines and so production of II is restricted. In this way we are faced with a problem of general equilibrium in which the evaluation of resources is determined not only by the prices of the products that they are used to produce but by the prices of competing products as well.

The restrictive condition of an isomorphic plain can easily be relaxed. Fig. 6.4 shows an increase in the marginal rent line between M_1 and M_2 as a result, for example, of a pocket of fertile soil. Alternatively it could result from a spatially selective government-financed inducement, grant, subsidy or price support designed to ameliorate the effects of

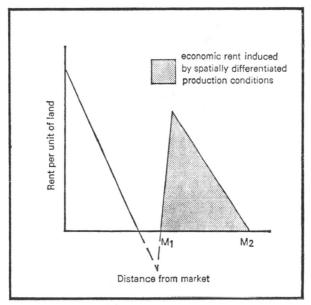

6.4 *Spatial variations and rent.*

economic marginality by encouraging production from marginal areas and having the consequence, like all price rises or cost falls, of increasing the value of the resources used to produce these products.

This discussion, based on an extension (Dunn 1954) of the analysis of von Thünen (Hall 1966), is highly simplified (Found 1972) and does not consider the similar process of land allocation in urban areas (Alonso 1964) but it reveals several important points about resource evaluation in general. First, the distinction between resources and reserves depends, in a capitalist economy, upon the distinction between positive and negative marginal returns resulting from their use. Secondly, this distinction is dynamic, changing in response to product prices and production costs. Both of these conclusions illustrate the concept of resource creation and destruction by the operation of the economy and they may be generalized to cover resources other than geographical space. For example, fig. 6.5b presents a more comprehensive taxonomy of natural resources as classified by the processes shown in fig. 6.5a. Demand and supply influences, reflecting a variety

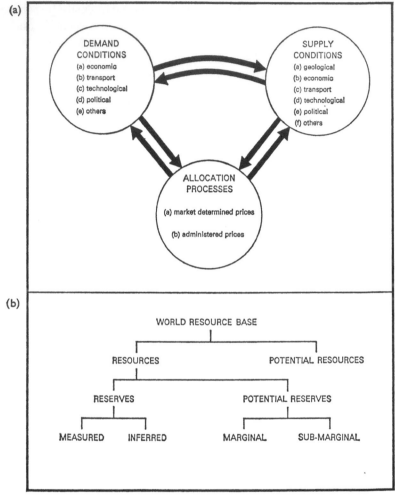

6.5 *Evaluation and taxonomy of resources.* (*Source: Manners 1969.*)

of factors, are reconciled by an allocation process which may result from the operation of a completely free market at one extreme to a completely administered process at the other. Clearly the distribution of reserves, as defined above, is a function not only of their demand and supply characteristics but also of the way in which these influences are measured and valued by the allocation process.

The capitalist market mechanism distributes income to resource owners in a socially arbitrary fashion. Landowners who happen to control land near to the town can earn higher incomes, whilst those further out earn a much lower income and those beyond the margin of land reserves earn nothing within the commercial economy; their resource of land remains unemployed. Such a distribution of income would be excused by the notion of marginal productivity which recommends that returns to factors be equal to the value of the marginal product produced by them. Of course this system of rewards takes no account of need and so tends to intensify existing inequalities of income. But, when applied to conditions in which resource owners are assumed to be completely free to move from one occupation or location to another, it suggests that the incomes of resource owners would be equalized. If owners offer their resources to those firms offering a high price, then production from these firms will increase, the price of their product falls and so the marginal value of a unit of resource also falls. By contrast, if owners withdraw supplies of resource from firms offering low prices, their production will fall, the price of their goods will rise and so increase the marginal value of the resources used by such firms. But this process can only ensure the equalization of the incomes of employed resources.

Spatial variations in income
A vital assumption underlying such instantaneous adjustment is the absence of movement costs so that resources are completely mobile. In the example discussed above landowners are unable to increase the supply of land to high-rent users in order to equalize incomes and, as we have already seen, labour and capital resources are not perfectly mobile. Thus locationally predictable but socially arbitrary inequalities result from the capitalist evaluation of resources within a space-economy. These inequalities are known as equilibrium differentials because they occur even when the demand and supply for resources are in equilibrium. Such inequalities in the distribution of income are often distinguished from dynamic differentials which result from changes in the derived demand for resources. When demand for the products of a given industry falls, one effect will be to lower the incomes to resources operating within that industry, because the marginal return of the products that they can be used to produce will also fall, if the product price drops in response to the fall in demand. Under conditions of resource mobility it might be expected that such dynamic differentials would be evened out as resources move from low-return to high-return occupations.

Spatial inequalities in income may result from equilibrium differences in that resources are spatially differentiated (fig. 6.5). But the fundamental determinants of income distribution in capitalist economies are the market mechanism, as applied to resource evaluation, and the distribution of resource ownership. Neither property in the means of production nor the ability to sell resources is evenly distributed. Yet transfer payments, such as old age pensions, unemployment pay, social security payments and redistributive income tax schemes, ameliorate the worst excesses of income inequality without removing private property and the market mechanism, which together ensure inequality. Such schemes treat symptoms rather than causes.

Descriptions of international inequalities in income are legion within the geographical literature alone (e.g. Fryer 1965; Ginsburg 1961; Warntz 1965) and the quantity of descriptive material, not only of income disparities but of the complex of associated inequalities (Berry 1960), provides the empirical data input for the traditional analysis of the economics of development (chapter 10).

Description and analysis of income variations at a more local scale are, curiously, less abundant considering the importance of income as a major social and economic variable.[1] This importance should be clear from comments made in chapter 4 about the influence of income upon the economic and social status of households. In a different context Burghardt (1972) has suggested that a measure of income per unit area would approximate the evaluation that society has made of differing locations, but the relevance of this point for economic geography was explored more fully in the previous section. Descriptions of spatial variations in income in the United Kingdom (Coates and Rawstron 1971) reveal a marked tendency for total income per capita to rise towards the south and east of the country, but such descriptions also suggest that this tendency is more marked for investment income (received for selling capital into the economy) than for other personal income. In the United States the incidence of poverty (measured very largely by the income variable) is predominantly located in the south (Morrill and Wohlenberg 1971). The attempt to explain this distribution of poverty incorporates environmental factors (resource endowment), distance and accessibility to major urban centres, social characteristics of the population and local labour market conditions. Morrill and Wohlenberg conclude that spatial factors account for the large area of poverty in those non-metropolitan regions of the United States unable to attract or compete with modern growth industries in the national economy. Clearly this economic evaluation of resources is space-selective. However, in the metropolitan areas of the country, which account for the largest numbers of poor, the problem is largely one of economic and social structure. Racial discrimination reduces the resource income of blacks whilst low levels of educational achievement

[1] The wider field of analysis in regional economic development has, of course, grown very rapidly in recent years (see chapters 2 and 10).

and high proportions of elderly people represent little of value to a competitive, capitalist economy. Similarly, the lack of firms paying high wages in the metropolitan areas reduces the overall level of demand for the goods and services produced by the resources located there.

Conclusions

The economic geography of resources is concerned predominantly with the influence of space and location upon the economics of resource evaluation. This point of view is well exemplified by Manners' (1971a) study of the world market for iron ore in which the operation of the forces of demand, supply and market allocation are most convincingly demonstrated. In this chapter we have attempted to go a little further to show that the process of resource evaluation has implications beyond those of allocative decision making. The market system of evaluation produces marked discontinuities in the distribution of productive resources and hence in levels of income and opportunities for economic development. Thus the increasing inequalities in levels of international development relate to the low value placed upon the resource base of the less-developed countries by the evaluative judgments of the world economy. But, as we have already pointed out (pp. 54–5), the work situation affects more than the distribution of income alone and its effects are apparent at a variety of spatial scales. Thus Smith (1973a) has taken the study of economic geography one stage further in an analysis of spatial inequalities in social well-being, incorporating income and a wide range of associated social indicators, at regional, inter-urban and intra-urban scales of analysis in the United States. The creation and destruction of resources results in changing spatial patterns of resource distribution (e.g. Chisholm and Oeppen 1973) and trade. But at the same time, resource evaluation is a major determinant of the distribution of income and of spatial inequalities in levels of social and economic development.

Part 3
The economy: interaction

The existence of space demands movement between the functionally specialized elements of an economy. But space also shapes movement because the friction of distance involves the costly allocation of resources to transport facilities and to the organization of flows through a system of markets. In this section we will consider the nature and significance of movement within economies and the role of market and transport networks in the physical integration of space economies and the satisfaction of movement demands.

7 Movement

Movement is an intrinsic feature of all economies. Even the un-specialized production of agricultural commodities within a spatially restricted subsistence economy involves movement. Labour must move to be combined with the resources of land, a little capital and the skill of the agriculturalist before production can take place, whilst the harvested crop must move to consumers before demands can be satisfied. These are small-scale movements but they are not spatially insignificant. The effort involved in making daily movements of this kind frequently results in spatial arrangements of agricultural production clearly designed to reduce the energy costs of movement. A detailed demonstration of distance effects upon the spatial and agronomic organization of production in four Indian villages has been provided by Blaikie (1971a and b), whilst Chisholm (1968) has extended this type of analysis to the world scale, and incorporated commercial agriculture, in a convincing study of the effect of movement costs upon the spatial organization of agricultural production.

The ability to move is also an important determinant of locational specialization. One cause of subsistence production over large areas is the limited access to markets provided by inadequate transport networks and storage facilities which together remove any incentive to produce a surplus for sale to distant consumers. Economic movement is clearly both cause and consequence of the spatial structure of economies (e.g. Törnqvist 1970) although in this chapter our attention will be limited, in the main, to a study of the causes of movement.

Classification

Several types of movement or communication are vital for the successful operation of an economy (fig. 1.1): (i) The communication of information about demands from consumers to firms and from firms to resource owners. (ii) The movement of resources and goods from the points of supply to points of demand or consumption. (iii) The complementary and opposite flow of money or payment in exchange for resources and goods received. (iv) The transmission, or diffusion, of innovations around an economy. This last type represents a rather different category of movement. However, ideas may be regarded as resources originating spontaneously or by design in households or firms and communicated through the economy by means of the spoken or written word and by the mass media.

These economically induced interactions are a major sub-set of more generally defined forms of human interaction (e.g. Haggett 1965; Abler, Adams and Gould 1971), but this chapter is concerned only with trade and money flows. The transmission of economic information and the movement of consumers in searching for goods has already been discussed whilst the study of innovation diffusion is beyond the scope of this book and forms a specialized field of enquiry (e.g. Hägerstrand 1967; Jones 1967). Trade flows may be defined as flows of resources or goods from points of production or processing to points of consumption or further processing – that is, between points of supply and points of demand. Money flows are generated in payment for traded goods and so flow in the opposite direction. The movement of both trade and money (henceforward simply referred to as 'trade') may be measured as flows between the economic sectors of an economy or between its constituent spatial units.

Sectoral and spatial movements

A feature of developed economies is that sectoral distinctions are minute because the occupational specialization of the elements in such economies is highly developed. As a result the movement of goods, information and ideas between sectors is both great in quantity and complex in structure.

Input–output analysis attempts to describe and measure the degree of sectoral interdependence within a national or sub-national economy and to assess the impact of disturbances, such as changes or shifts in final demand, upon the structure of an economy's flows, output and income. These tasks are accomplished with the use of input–output tables or matrices (table 7.1) which divide an economy into a number of sectors listed as the headings of both columns and rows. The rows represent the sectors as producers and include a row or rows for resource-owning households and imports, whilst the columns represent the sectors as consumers and include columns for final demand by consumers, private investment and government spending, and a column for exports. The cells of the matrix contain the value of the transactions between producing and consuming sectors so that the way in which the output of a sector spreads through the economy and its contribution to total output can be seen from the row-cells. The input to a sector and the variety of inputs that it needs may be seen by scanning the appropriate column.

The main analytical purpose of input–output tables is to determine the effects of specified changes in final demand upon gross output from each producing sector. Certain assumptions are necessary for predictive purposes, including the most critical – that industrial output per unit of

TABLE 7.1 *Simplified input–output table*

From	To	Purchasing sectors 1 ...j ...n	Local final demand Households C	Private invest- ment I	Govern- ment G	Ex- ports	Total gross out- put
	1	$X_{11}...\ X_{1j}\ ...\ X_{1n}$	C_1	I_1	G_1	E_1	X_1
		. . . .	.	.	.	.	.
		. . . .	.	.	.	.	.
Pro-ducing sectors	.	. . . .	.	.	.	.	.
	i	$X_{i1}\ ...\ X_{ij}\ ...\ X_{in}$	C_i	I_i	G_i	E_i	X_i
		. . . .	.	.	.	.	.
		. . . .	.	.	.	.	.
		. . . .	.	.	.	.	.
	n	$X_{r1}...\ X_{nj}\ ...\ X_{nn}$	C_n	I_n	G_n	E_n	X_n
Labour		$L_1\ ...\ L_j\ ...L_n$	L_C	L_I	L_G	L_E	L
Other value added		$V_1...\ V_j\ ...V_n$	V_C	V_I	V_G	V_E	V
Imports		$M_1\ ...M_j\ ...M_n$	M_C	M_I	M_G	—	M
Total gross outlay		$X_1\ ...\ X_j\ ...\ X_n$	C	I	G	E	X

Source: Richardson (1972)

input remains constant whatever volume of output is produced (Richardson 1972). The effects of changes in final demand not only include the first-round direct impact on the economy but also successive rounds of the indirect effect of additional intermediate demands upon all or most sectors. Input–output tables are descriptive of both the sectoral structure of economies and the movement within them and they provide a powerful analytical and predictive tool with which to explore the sectoral effects of an increase in economic activity and the sectoral location of potential bottlenecks for economic expansion. They may be expanded to include several regions so that aggregated spatial changes resulting from sectoral adjustment may be predicted. However, the problems of data shortage and an increase in the openness of economies, as spatial detail is increased, complicate the regional extension of input–output tables.

Economic movements are also measured as flows within and between places, each of which may contain several representatives of all the elements of an economy, operating within several sectors, so that some of the detailed structure of movement described by input–output matrices is lost. The use of areal units as basic units of measurement is common to all scales of trade-flow analysis (table 7.2) but the implications of the use of these areal units are more complex than simple scale

TABLE 7.2 *Areal units for movement analysis*

Scale of movement	Movement measured between
Intra urban	Traffic zones
Inter urban	Urban nodes
Inter regional	Traffic regions
International	Nation states

effects. The extent of internal homogeneity and the degree of openness of the areal units, which affect the amount of internal movement generated and the ease with which external movement can take place, vary inversely with their size and political significance. Thus national and supra-national units have a high degree of internal heterogeneity and closure (Keeble 1967). One consequence of this is that although interaction is greater between smaller areal units, trade statistics are more readily available for larger areas and nation states, since political boundaries provide an efficient filter for the collection of data. Thus the United Nations publishes annually a set of statistics on the volume and value of international trade movements to which 142 countries, accounting for 98% of world trade (in 1968), contribute information. Despite this relative abundance of data and the well-developed branch of international economics, geographers have not paid a great deal of attention to international trade (Thoman and Conklin 1967; McConnell 1970; Johnston 1973). But several studies of movement at a subnational scale provide insights into the spatial structure of economies and the bases of spatial interactions (Ullman 1957; Berry 1966; Chisholm and O'Sullivan 1973) and so go well beyond the more limited studies of international trade in exploring the connections between relative locations, locational specialization and the attributes of movement.

A strategy for the spatial analysis of movement

During recent decades the intense problems of movement in urban areas have become increasingly apparent (e.g. Ministry of Transport 1963; Garrison 1966), but attempts to deal with the problems were hampered until the early nineteen-fifties by the lack of an appropriate conceptual framework for the study of movement. Previous studies had been based upon roadside counts of traffic flow and surveys of published transport time-tables which were then extrapolated to some future date with the application of an appropriate growth factor. This method of study clearly analyses effects rather than causes, but in 1954 Mitchell and Rapkin shifted the emphasis towards the analysis of cause and effect. Using data from the metropolitan area of Philadelphia they showed that traffic generation and flow were closely related to the pattern of land use in urban areas. This conceptual breakthrough gave rise to the numerous Urban Land Use Transportation Studies (Zettel and Carll 1962), employed first by many North American cities and, with central government encouragement since 1964, by metropolitan local authorities within the United Kingdom.

Although urban transportation studies are conducted at a particular scale their concern is with the measurement and prediction of transport demand and the methodology developed by them is equally applicable to other scales of movement analysis. The traffic zones, used as the basic data-collecting units in urban studies, are only a small-scale example of the areal units used in all spatial analyses of movement (table 7.2). The analysis of urban movement proceeds in two major stages and is followed by a third evaluation stage. Berry and Horton (1970) provide a useful review of this strategy for intra-urban analyses and Johnston (1965) has edited a comprehensive case study of its application to Christchurch, New Zealand.

I SURVEY AND ANALYSIS

(i) The build-up of an inventory of current conditions affecting the present demand for movement. Data are collected for explanatory variables such as land use, population size and socio-economic structure, current availability of movement facilities and the budget allocation for their improvement.

(ii) The building of models of traffic generation and flow characteristics derived from the quantified relationships between the independent variables described in (i) and current levels of movement. In urban studies data on the latter are collected from roadside origin-destination surveys and from interviews with households and firms. As we have noted already, published information is more readily available for analyses of movement at larger, especially international, scales, although marked spatial contrasts exist in the availability of such data. In Britain, for example, there are no comprehensive data on interregional movements of freight whereas in the United States waybill data are collected by the Interstate Commerce Commission and in India Inland Trade Accounts are published by the Government Department of Commercial Intelligence and Statistics. Model building is normally concerned with the following movement characteristics: generation, by origin and destination; distribution or allocation (intensity of interaction between traffic zones origins and destinations); assignment (to routes between interacting zones, noting the predicted level of utilization on each link of the transport network); modal split (share of movement taken by alternative modes of transport).

II PREDICTION

(i) The prediction of change in the explanatory variables to some future date is made with the use of exogenous models (e.g. models of population change) and must take into account political influences upon change.

(ii) The estimated future values of the explanatory variables are then used within the traffic models to predict the various attributes of traffic generation and flow.

Studies of the demand for movement in urban areas provide data for decisions about investment in transport facilities. Such a topic is more logically considered in the chapter on transport, but as this evaluation process is the vital third stage of urban transportation studies we introduce it here.

III EVALUATION AND PLAN FORMULATION

(i) The predicted demands for movement are matched with alternative transport investment strategies.

(ii) Given the physical and economic viability of using the proposed networks and the budget constraint of available investment funds for building them, a choice must be made between the alternative investment proposals. The choice of positive action lies, theoretically, along a continuum, with the complete rebuilding and restructuring of the outdated transport system to meet predicted demands at one extreme and a limited modification of existing networks combined with the rigid control of movement at the other.

The structure of spatial interaction

This brief description of the methodology of urban transportation studies provides us with a logical and structured model for the study of movement in space. In this section we begin by considering traffic generation (the demand for movement in space) and move on later to consider some of its spatial or distributional characteristics.

THE GENERATION OF MOVEMENT WITHIN AREAS

The total volume of movement within an area is a function of the size and composition of its economic output (e.g. G.N.P. or G.D.P.), the degree of integration between the elements of its economy and its areal size and shape. The history of international trade (e.g. Lamartine–Yates 1959; Courtenay 1972; Paterson 1972) shows that this generalization is broadly acceptable for a time-series view of movement generation at the world scale. World trading networks have evolved during the present century, from a simple colonially-based system to a much more extensive and complex structure. This change has been wrought by factors like the increase in the number of non-European economies participating more fully in world trade; the post-war emergence of Japan as a major origin and destination of trade and of petroleum as the most voluminous commodity, introducing many new nations into the world trading system; the growth to maturity of the United States' economy during the latter half of the nineteenth century and its emergence as a world power in the twentieth; the rapid and, in the case of many countries in Western Europe, sustained growth of the advanced capitalist and communist economies during the post-war period. With the exception of the disruptive effects of the Great Depression and two World Wars, these changes have been accompanied by a growth in the volume of trade generated. However, table 7.3 shows that this growth, whilst dramatic, has been uneven.

TABLE 7.3

Growth in the value of international trade 1938–74 ($million)

Region	1938	1948	1958	1968	1974	Percentage increase 1938–74	1968–74
Developed market economies	33 000	77 800	145 500	348 100	1 155 600	3 401·8	232·0
Less-developed market economies[1]	10 000	30 100	39 700	67 100	222 100	2 121·0	231·0
Centrally planned economies	3 300	7 400	25 100	55 000	151 700	4 497·0	175·8
World	46 300	115 300	210 300	470 200	1 529 400	3 303·2	225·2

Source: Derived from United Nations (1976); [1] excluding O.P.E.C. (see p. 82)

Integration in the world economy is by no means complete and so acts as a constraint upon the demand for movement. The level of development within an economy is an important variable affecting its integration, and the spatially selective processes of economic growth (chapter 10) continue to provide one very important constraint upon the future growth of world trade. Many communities within the less-developed countries have few connections outside their own immediate locality and it would appear that in relative terms at least, such countries are being pushed further and further away from the main stream of economic interaction within the world economy (table 7.4). The influence of political and social ideology has also produced international inequalities in socio-economic systems which limit economic integration and tend to reduce the total volume of world trade, as conscious attempts at self-sufficiency are often part of the policy of centrally planned

TABLE 7.4 *Regional shares of world trade 1938–74*

Region	1938 %	1948 %	1958 %	1968 %	1974 %
Developed market economies	71·3	67·5	69·2	74·0	75·5
Less-developed market economies[1]	21·6	26·1	18·9	14·3	14·5
Centrally planned economies	7·1	6·4	11·9	11·7	10·0
World	100·0	100·0	100·0	100·0	100·0

Source: Derived from United Nations (1976); [1] excluding O.P.E.C.

economies. Certainly these economies account for a much lower proportion of international trade than their productive power would suggest (tables 7.4 and 1.1).

The fact that international trade crosses national political boundaries facilitates its regulation by governments. Decisions on trade, taken in the first instance by individual firms and households, are modified in their effect by a complex system of institutional arrangements having several aims and methods of application (e.g. table 7.5; Thoman and Conklin 1967). During the twentieth century two periods, the first during the years of the Great Depression between 1929 and 1933, and the second during the emergence of the major world economies from the Second World War, have seen the erection of restrictive institutional barriers to trade. The General Agreement on Tariffs and Trade (G.A.T.T.), which began operations in 1948, has been one reaction to this latter period of restriction. Its 76 members are pledged to the expansion of multilateral trade and the reduction of barriers to trade. The increase in free trade is to be achieved by successive rounds of negotiations. How-

TABLE 7.5 *Institutional trading arrangements*

Form	Purpose
A. National arrangements	
Trade promotion ⎫ Export subsidies ⎭	Stimulation of export trade
Tariff barriers	Taxation of imports
Import quotas	Quantitative restrictions on imports
B. International arrangements	
Commodity agreements	Regulation of production and trade in commodities
Free Trade Area	Promotion of trade within international groupings of states by removal of tariffs, quotas, subsidies etc.
Customs Union	As with Free Trade Area but also incorporating a common external tariff
Common Market	As with Customs Union, but also incorporating free movement of factors of production
Economic Union	Harmonization of economic policy and activity within international groupings of states
General Agreement on Tariffs and Trade (G.A.T.T.)	Promotion of multilateral international trade by removal of institutional restrictions
United Nations Conference on Trade and Development (U.N.C.T.A.D.)	Promotion of export trade from less-developed countries and protection of their terms of trade

ever, the members of G.A.T.T. reserve the right to reimpose barriers under conditions which may threaten national economic health.

In 1964, the first United Nations Conference on Trade and Development (U.N.C.T.A.D.) was held with the general intention of developing a united policy amongst less-developed countries in the attempt, for example, to persuade developed economies to give preferential tariff-treatment to manufactured exports from the poorer countries. The practical success of U.N.C.T.A.D., which represents an attempt to increase world trade on terms which will offer positive discrimination to the less-developed majority, has been limited, partly because of the divergent interests of the less-developed countries in acting as a single body and partly because of protectionist attitudes in many advanced economies supporting the imperialism of multi-national firms.

Another major influence upon international integration, and hence upon the demand for movement, has been the geographically more restricted but economically more comprehensive attempts to integrate the economic activities of groups of national economies, the most successful example of which is the European Economic Community. Several forms of economic integration may be recognized (table 7.5), but a common feature is the removal of restrictions upon trade between the nations involved. A customs union involves both a common external tariff against goods entering from third countries and the removal of internal barriers to trade. Fig. 7.1 illustrates the implications of a customs union for trade. Trading conditions before the union of

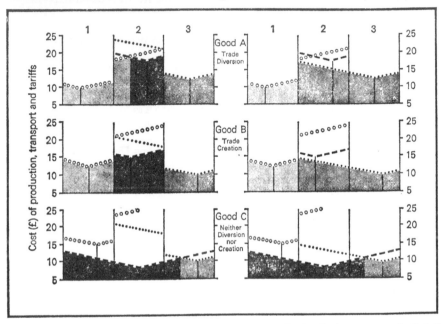

7.1 *Trade diversion and creation. (Source: derived from a tabulated example in Swann 1975.)*

countries 2 and 3 are shown in the left-hand diagrams and conditions after union in the right-hand diagrams. *Trade diversion* is said to occur in the case of good A as the common external tariff around the customs union blocks goods from the lowest-cost producer of A – country 1. *Trade is created* for good B as the lowest cost producer of B – country 3 is able to extend its market area within the customs union. Neither diversion not creation takes place in the case of good C.

This discussion of the factors affecting the total volume of trade generated within the world economy has been couched in very general terms. A more precise quantitative attempt to analyse the effect of the size of economic output upon the demand for movement within a national economy has been made in a study of the generation of freight traffic in Great Britain. Noting the surprising sparsity and statistical inadequacy of previous attempts to relate the growth in freight traffic to other indicators of economic growth, Chisholm (1971b) analysed the relationship between national aggregates of freight tonnage and ton-miles and the size of the G.D.P. over the period 1953–68. The results were disappointingly inconclusive (table 7.6) and indicate that changes in the future volume of freight cannot readily be predicted from changes in G.D.P. Furthermore, Chisholm found that the simple relationship

TABLE 7.6 *Regression equations for freight traffic –*
Great Britain 1953–8 ($Y=a+bX$)

Y	$X = $ *Gross Domestic Product at 1963 Prices*				
	a	b	R	R²	*Students* T, *level of probability*
1. All freight, million tons	8·119	0·050	0·536	0·287	0·975
2. Road freight, million tons	26·312	0·028	0·305	0·093	0·800
3. Rail freight, million tons	−20·154	0·022	0·772	0·596	0·995
4. All freight, million ton-miles	−0·661	0·003	0·747	0·558	0·995
5. Road freight, million ton–miles	0·784	0·001	0·446	0·199	0·950
6. Rail freight, million ton–miles	−1·101	0·002	0·721	0·520	0·995

Note: all units are Imperial
Source: Chisholm (1971a)

between G.D.P. and freight per head of population or employed population cannot be sustained with time as an independent variable. These results, combined with the lack of suitably detailed and comparable data, restrict the extension of such time-series analyses to comparative cross-sectional studies of space economies. Furthermore, in the case of the British economy, aggregate results at the national scale cannot be linked with procedures for allocating freight to a spatially disaggregated system of regions (Chisholm and O'Sullivan 1973). This is due not only to the poor performance of attempts to predict traffic generation at the national level but also to the absence of regional data on national income.

THE GENERATION OF MOVEMENT BETWEEN AREAS
Movement or trade between economies takes place because spatial patterns of demand do not correspond with the spatial availability of supply. Both spatial and non-spatial factors underlie the generation of this movement. The spatial factors relate to the influence of relative

location upon the generation of movement, whilst the non-spatial factors comprise the economic characteristics of the places generating the movement. However, although it is normal to consider trade as flowing between areas, it must be remembered that many of the decisions about trade are taken by individual firms and households within the areas. In the case of consumer goods spatial discrepancies of demand and supply are caused in part by consumer behaviour and the locational behaviour of firms. These discrepancies may provide firms with the opportunity of selling goods abroad or of importing foreign goods for sale on the home market. The success of these ventures depends upon consumers' reaction: if they find the goods attractive they will buy them and the trade will continue; if not the trade will cease. Despite the fundamental importance of individual decision making, economic studies of trade conventionally consider that the economies as a whole, rather than the individual elements, are the basic trading units. Clearly, a behavioural approach to trade, especially at the international scale, is a potentially very valuable field of study.

Non-spatial factors
Non-spatial factors affecting the generation of movement have been most intensively studied at the intra-urban scale. The large number of urban land use transportation studies have revealed that three variables (income, car ownership and household structure) are responsible for much of the inter-zonal variation in the domestic generation of movement and that the size and composition of economic activity are clearly discernible zonal influences upon non-residential traffic generation (e.g. Eliot-Hurst 1970) which may also be influenced by the spatial factor of zonal location relative to the C.B.D. (e.g. Taylor 1968). At a more detailed level, Starkie (1967) used individual manufacturing plants rather than traffic zones as the basic unit of observation in a study of industrial traffic generation in the Medway towns of Kent. His results showed that the relationship between movement generation and size of unit was curvilinear. This form of relationship is explicable in terms of transport economies of scale in larger plants and so in their study of traffic generation within 78 traffic zones of the British economy Chisholm and O'Sullivan (1973) assumed a linear relationship between the dependent variable of the volume of traffic generated and the independent explanatory variables. The attempt to explain movement generation at this scale was concerned with the influence of the level of economic activity within each traffic zone. The variables chosen to act as surrogates for this were total population, total employment and retail turnover. For road and rail freight taken together, these independent variables accounted for between 65% and 75% of total variance, whereas for road alone 74% to 79% of the variance was ex-

plained. The results for rail alone were poor because rail freight is dominated by mineral and bulk goods traffic emanating from a limited number of origins and consigned to a few major points of consumption. A disaggregation of total freight by commodity did not improve the level of explanation although when the employment structure of each zone was disaggregated into 24 orders of the Standard Industrial Classification the level of explanation for many commodities reached 80% to 90%. Again road and rail considered together fared less well than road alone. A conclusion reached by this study was that the results were not good enough to permit confident forecasting. However, the use of readily accessible employment and population data could allow a series of simulation exercises to establish the future range in levels of traffic generation, given forecasts and assumptions about the future geography of population and employment.

A qualitative description of some of the non-spatial factors affecting the generation of movement by economies at an international scale has been provided by Kindleberger (1963), whilst Linnemann (1966) suggests that the concept of potential foreign trade underlies these non-spatial factors. An economy's potential trade supply or demand may be defined as that part of its total output or consumption not designed for or oriented to domestic consumption or production, assuming that the resistance to trade in the international market is the same for all economies. This assumption removes the influence of specific and differential factors upon the trade of particular countries and points to the determinants of the ratio between volume of output for the domestic and foreign markets (the DM/FM ratio) as being of fundamental significance for the generation of foreign trade. The value of this ratio is linked systematically to: (a) the size of an economy's total output which is included only as a scale factor because the size of exportable product is scarcely influenced by the size of G.N.P.; (b) the size of an economy's population because a large population increases both its economies of scale and self-sufficiency and so enlarges the DM/FM ratio; (c) the per-capita income of the economy which has only a limited effect because it increases both the numerator and denominator of the DM/FM ratio. These hypotheses were confirmed in a regression analysis of eighty countries for the year 1959 which showed that G.N.P. was a major influence upon the volume of trade generated but that this effect is reduced by larger populations and hardly affected by the level of income per head (see p. 130).

Spatial factors
The existence of geographical space, and the unequal distribution of economic activities in space, affect the demand for movement by introducing locational relationships between the interacting spatial

units. Ullman (1956) proposed a three-factor system, based upon the notions of complementarity, intervening opportunity and transferability, to describe these locational influences and so to explain the demand for movement between places.

(a) Complementarity. Economic movements between areas are a function of the specific complementarity between them; demands in one area must be capable of being met by a supply in the other. The concept of complementarity forms the basis of economic base studies (Dziewonski and Jerczynski 1971; Abler, Adams and Gould 1971), and the opportunity cost or comparative advantage approach to regional (e.g. Smith 1964b) and international trade (Wells 1969). We shall here consider this latter application in more detail.

In fig. 7.2a a production possibility curve is shown for an economy

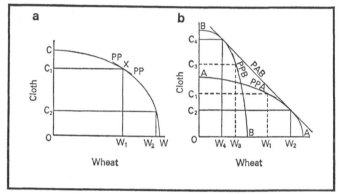

7.2 *The gains from local specialization. (Source: Wells 1969.)*

producing cloth and wheat. Such a curve shows the amount of wheat and cloth that can be produced with the resources at the economy's disposal. It can produce either OC units of cloth and no wheat, or OW. units of wheat and no cloth, or any combination of the two products, for example, OC_2 units of cloth and OW_2 units of wheat. The actual combination produced will depend upon the price ratio of the two commodities. This ratio is shown in fig. 7.2a by the price line PP. It intersects the production possibility curve at X and this combination of production, would, under competitive conditions, provide no incentive for producers to change their output. The reasons for this were discussed in chapter 5. Under conditions of perfect competition output can only be in equilibrium where price (marginal revenue) is equal to marginal and average costs. Thus, at equilibrium levels of output, the price ratios of two commodities must reflect their marginal cost ratios so that revenue cannot be increased by adjusting the combinations of output.

In fig. 7.2b the production possibility curves for two countries A and B are shown. In B conditions favour the production of cloth rather than wheat, whilst the reserve is true in country A. The equilibrium production points, where the price lines in each country (PPA and PPB) touch their production possibility curves, illustrate the differences in the opportunity cost between A and B. In A the opportunity cost of increasing the output of cloth would be expensive in terms of the output of wheat that must be foregone; by contrast any attempt to increase wheat production in B would result in an expensive loss of cloth. In short, A has a comparative advantage in the production of wheat, B in the production of cloth.

The demand for interaction between the two countries stems from their differential opportunity costs. As wheat is expensive of cloth in B (its opportunity cost is high) and cloth expensive of wheat in A (its opportunity cost is also high) both countries could benefit by expanding their output of the low-cost product, thereby allocating resources to the production jobs they are technically best fitted to perform, as illustrated by the different shape of the production possibility curve in each country. The implication is that by locational specialization and trade both countries could be made better off as more of both commodities can be produced.

The locational specialization that follows the opportunity to trade is shown in A by the expansion of wheat production from OW_1 to OW_2 and the decline in cloth output from OC_1 to OC_2. It is shown in B by the expansion of cloth production from OC_3 to OC_4 and the decline in wheat from OW_3 to OW_4. As trade commences there can only be one price ratio for the two commodities. This is represented by PAB in fig. 7.2b and the new price ratio shows that as the imported supply of cheap wheat increases in B its price falls, the old price line in B decreases its slope whilst the import of cloth in A lowers its price and increases the slope of its price line. Trade continues until all the benefits of locational specialization are exhausted – that is until the price ratios in the two countries are equated. Thus A exports W_4W_2 units of wheat to B and imports C_2C_4 units of cloth from B.

The concept of comparative advantage, based upon spatial differences in opportunity costs, elegantly explains the derivation of the demand for movement between areas but it offers no suggestion as to how or why the differences arise. Ullman (1956) suggests that complementarity is a function of natural and cultural areal differentation and the locational specialization of economic activities. According to Ohlin (1935) spatial variations in factor endowments are the most significant cause of variations in opportunity cost and hence of the development of complementary links between places and the growth of trade. Ohlin's assumption, that production functions (the technical relationship between inputs and outputs) are the same in each place engaging in trade, clearly places variations in factor endowments at the centre of the explanation for trade flows by removing any chance of *in situ* production adjustments to factor endowments. Under this assumption most individual economies could obtain many goods at reasonable cost only by engaging in trade because the costs of domestic production under conditions of severe factor scarcity would be prohibitive. However, trade encourages local economic specialization, allowing activities which need large amounts of a certain factor, to gravitate towards low-cost sources of that factor.

The apparently paradoxical conclusion that the United States economy engages in trade in order to take advantage of surplus labour and economize on scarce capital, by importing capital-intensive goods and exporting labour-intensive goods, was based upon this assumption (Leontieff 1954). In accepting it Leontieff showed that import-replacement industries were apparently more capital intensive than the export industries. However, because the United States is a capital-rich economy, both import-replacement and export industries are capital-intensive by comparison with the relatively poor capital endowment of her trading partners. Thus an implication of accepting the assumption that production functions do not vary in space is the addition of a large amount of spurious capital to the import replacement industries. For more accurate comparisons the capital intensity of United States imports should be examined in terms of the production conditions of the exporting countries rather than in terms of United States conditions.

The complementarity between places is affected by spatial variation in demand as well as supply. This fact was recognized by Ohlin (1935) who pointed out that differences in demand between two countries with similar factor endowments could stimulate trade by affecting product and hence factor prices. However, the concept of comparative advantage as an explanation of trading has been criticized by Linnemann (1966) on the grounds that its relative importance differs little from country to country, although it may be argued that similar market demands as opposed to different opportunity costs can engender trade between areas. A large home demand allows large-scale production, and similar demand structures between regions facilitate the easy penetration of export markets without costly adjustments to products to cater for the demand conditions in the export market. This hypothesis has been used as an explanation of the large amount of intra-European trade between areas with similar factor endowments and opportunity costs (Linder 1961) and could be extended to explain some of the difficulties experienced by the less-developed world in breaking into the world trading system.

So far in this discussion we have considered the concept of complementarity as a general factor underlying all interactions between areas, but, as we have already noted, Ullman (1956) took care to point out that specific complementarity is a pre-condition of movement between any two areas. A further pre-condition is also necessary: specific complementarity will not stimulate movement in a capitalist economy unless the interaction generates some form of economic gain. To return to our previous hypothetical example, country A would not specialize in the production of wheat and exchange the surplus for cloth unless the amount of cloth obtained is worth more than the amount which could be produced domestically as an alternative to producing the surplus wheat. The same condition holds in the case of the surplus production of cloth in country B. These minimum benefits from trade are known as the limiting rates of exchange; at any point beyond the limits trade is not worthwhile[1] and so the ability of complementarity to generate movement is diminished.

Within the limits set by the limiting rates of exchange, the ratio of the prices at which the goods are exchanged (the terms of trade) depends upon the demand of the two countries for each others' produce and the ability to expand supply in response to this demand. If the demand for cloth in country A is greater, relative to supply, than for wheat in country B, then the greater part of the gains from trade will accrue to B provided that B can expand the output of cloth quickly enough to meet the demand. An economy's barter or commodity terms of trade (T) is normally measured as a ratio of the price indices of exports and imports:

$$T = \frac{P_x}{P_m} \ 100$$

where P_x and P_m = price indices of exports and imports respectively.

The terms of trade improve if the prices of exports rise relative to those of imports and deteriorate if the prices of exports fall. The Prebisch thesis (Prebisch 1950) suggests that the long-run terms of trade are moving against the less-developed countries and reducing the quantity of imports obtainable with a given volume of exports. This tendency results from spatial inequalities in the gains from technical progress and increases in productivity. The imperfectly competitive manufacturing firms located predominantly in the developed countries are able to withhold output from the market and so prevent a fall in price. By contrast, the more nearly perfectly competitive conditions of primary producers are less able to prevent price reductions following increases in productivity. In this way the gains from technical progress accrue very largely to the developed countries (but see p. 82).

Furthermore, Prebisch (1950) suggests that the income-elasticity of

[1] A worked example of limiting rates of exchange is in Chisholm (1970, 19–22).

demand for primary products is less than 1 but for manufactured goods is greater than 1. This means that increases in income create a proportionately greater increase in the demand for manufactured goods and a smaller increase for primary goods (e.g. table 4.1). The low income-elasticity of demand for primary products stems from non-proportional reductions in consumer expenditure on food as incomes rise, and the increasing tendency to substitute more reliable, closely controlled and often superior manufactured commodities for more expensive and technically inferior natural products.

There have been several criticisms of these arguments, not the least of which is based upon the measurement of the terms of trade. A distinction should be drawn between (i) an improvement in the terms of trade which result from a rise in export prices stimulated by an increase in foreign demand and (ii) an apparent improvement caused by an increase in domestic costs. The factoral terms of trade (T_f) refines the simple barter measurement by incorporating a measure of productivity in the export sector:

$$T_f = \frac{P_x Z_x}{P_m}\ 100$$

where Z_x = an index of productivity

Changes in the factoral terms of trade can result either from domestic productivity changes or from price changes. Another refinement to the barter measure, known as the income terms of trade (T_i), takes account of the fact that a country's export income is a function of the product of the volume and price of exports:

$$T_i = \frac{P_x V_x}{P_m}\ 100$$

where V_x = an index of the volume of exports

The use of this measure shows that during the post-war period the terms of trade of both advanced economies and the primary exporting less-developed countries have improved, although the former improved much more rapidly than the latter. Thus a worsening of the barter terms of trade need not be accompanied by loss of national income and could be offset by improvements in productivity, although the capital-scarce less-developed countries are less able to increase productivity than are the advanced economies. Such a policy could, in any case, lead to large-scale unemployment of labour resources.

Changes in the terms of trade can have several significant effects upon trading partners (Chisholm 1970). First, they can redistribute income between the partners so that a deterioration in the commodity terms of trade, despite productivity improvements, implies a lower national income than if this deterioration had not occurred; secondly, an economy's balance of payments suffers with worsening terms of trade and in recent years this effect has been one of the most potent causes of government interference in trade and money flows; thirdly, the effects of reduced income may spread from the export sector to the rest of an integrated economy.

Short-term changes in the complementarity between trading partners may result from rapid fluctuations in the price of traded commodities. These fluctuations are frequently caused by inelastic demand and supply typified by the conditions facing primary products. The difficulty of controlling their supply arises from the large input of natural processes to their production, whilst they are also susceptible to cyclical demand fluctuations. Various international commodity agreements (e.g. coffee, sugar and tin) have been established to regulate such fluctuations, but for many reasons their effect has been limited and may even have increased price fluctuations. By providing a guaranteed market and price for a restricted volume of output, the agreements tempt increases in output both from producers in the countries forming part of the agreement and from those outside who are encouraged to produce by the higher, more stable prices. Conversely, consumers are normally willing to purchase supplies at prices below that agreed.

National control of short-term fluctuations induced by international changes in complementarity involve the measures listed in table 7.5 and, in less-developed countries, the growth of marketing boards with national monopoly control of buying and selling primary agricultural commodities. The effective operation of these boards is plagued by lack of data and by their inability to act as rapidly as the freely operating price mechanism. Furthermore their effectiveness is limited by the inability to control natural conditions which may in turn induce fluctuations in output and instability of incomes.

The emphasis upon international trade that has prevailed in the last few pages indicates that the terms upon which trade takes place are most easily measured and manipulated at the boundaries of nation states. But the problems are common to all scales of analysis, and the openness of sub-national regions is an important element in the generation and sustenance of regional inequalities, as regional terms of trade, incomes and balance of payments are affected by exactly the same types of transaction.

(b) Intervening opportunities. Complementarity can generate movement between areas only if no intervening source of supply is available or preferred. When the limiting rates of exchange are exceeded it becomes economically rational for a country to replace trade for the intervening opportunity of domestic production and to substitute goods produced at home for imported commodities. More generally, the attractive effects of a third area can reduce the significance of complementary relations between two potential trading partners. The trade-diversion produced by a customs union is an example of institutionally induced intervening opportunity, as the complementarity between two trading partners is disrupted by the erection of an external tariff barrier (fig. 7.1). The concept of intervening opportunity has implications for modelling the spatial distribution of flows and so will be considered again in the next section.

(c) Transferability. If the distance between demand and supply points is too costly, movement will not be generated and interaction will not take place, despite perfect complementarity within the limiting rates of exchange and no other forms of intervening opportunity. The influence of transferability upon the generation of movement is decreasing in importance and may be exemplified by the rapidly changing structure of international trade in iron ore during the post-war period (Manners 1971a).These changes have been facilitated by economies of scale in bulk transport and large-scale centralized buying in the importing nations. However, falling transport costs benefit those commodities (e.g. bulky raw materials and foodstuffs) and remote economies (e.g. Australia and New Zealand) on which transport costs bear most heavily. Together with the particularly rapid advances in bulk transport techniques and more recently in general cargo handling, the differential effect of reductions in the inhibiting effect of transferability upon movement may lead to changes in the terms of trade between countries exporting low value goods and those exporting high value goods, and between countries at different distances from the major centres of international supply and demand (Chisholm 1970).

The factors of intervening opportunity and transferability operate in a spatially selective manner to inhibit movement generated by the complementarity of areas. Location theory suggests that an important influence upon spatial variations in the economic gain to be derived from interaction between areas (the factor of complementarity) is their relative location within a space economy. Transport costs are a function of distance. One way of reducing them, in the cause of increasing profits, is for firms to locate close to markets and suppliers. A location in proximity to many other firms may also give rise to external economies of scale and, for many industries, provide easy access to both consumers and suppliers. All these factors contribute towards the development of a concentrated central area and a remote, less economic, periphery as firms tend to choose locations central to the space-economy in order to minimize transport inputs and maximize economies of scale.

The spatial adjustments made by firms to these variations in locational advantages may include a reduction in the volume of movement generated in remote areas in response to the increase in intervening opportunities, coupled with a lengthening of the mean distance of trips generated within the more remote areas. By contrast, areas within the centre of an economy may take full advantage of a short mean length of haul, representing a high level of transferability, by engaging more fully in spatial interaction as a result of the reduction in intervening opportunity. Chisholm and O'Sullivan (1973) tested the first of these hypotheses for the 78 traffic zones within the British economy by analyzing the residuals from the regression of total road freight upon total population ($R^2 = 0.77$, for origins; 0.79, for destinations). If the peripheral locations engage in economic activities generating a low freight tonnage, as one means of avoiding the disadvantages of remoteness, the regression equation may be expected to overestimate freight generated at the periphery and underestimate it at the centre. However, the residuals were found to be completely unrelated to potential accessibility and they showed that in this case at least, the notion of centrality is irrelevant to this aspect of the generation of movement between areas.

At an international scale Linnemann (1966) points out that the potential foreign trade of an economy (see above p. 115) is inhibited by a distance variable incorporating natural (e.g. transport cost) and artificial (e.g. trade control) restrictions upon movement. An analysis of the foreign trade proportions (measured as a ratio between the average value of imports plus exports and G.N.P.) in ten differing countries showed that G.N.P. operates mainly as a scale variable, and that although the explanation of variations in foreign trade proportions is complex the major influences are the population size of the economies and the distance separating them from their trading partners, a conclusion also reached by Beckerman (1956) in a study of intra-European trade.

Ullman (1956) points out that intervening opportunity results in the substitution of areas as it affects the choice of trading partners whilst transferability, in so far as it rules out trade between areas, results in a substitution of products as domestically produced goods must replace imported goods. However these forms of substitution are not mutually exclusive and product substitution may accompany substitution between places.

THE DISTRIBUTION (OR ALLOCATION) OF MOVEMENT

Having investigated the total volume of movement generated by an economy and its constituent spatial units, the next stage is to analyse the distribution, or allocation, of the total movement between the interacting pairs of origins and destinations (known as dyads) within the space economy. This allocation is not necessarily concerned with the mode of transport used or the precise route taken by the flows; its main purpose is to pair the origins and destinations of trips. Four commonly used approaches to this task of allocation – gravity models, intervening opportunity formulations, multiple regression analyses and linear programming – will be briefly described. This will be followed by a short discussion of the spatially complex structure of commodity flows induced by physical handling problems.

The gravity model

The gravity model incorporates concepts from social physics in a simple relationship demanding only a limited data input. It states that the interaction between two places is directly proportional to their size or mass and inversely proportional to the distance between them. This may be expressed as:

$$I_{ij} = K\frac{M_i M_j}{D_{ij}}$$

where I_{ij} = interaction between places i and j
       M_i, M_j = size or mass of places i and j
       D_{ij} = distance between i and j
       K   = a constant

The constant (K) is necessary in the empirical application of the model to ensure the equivalence of the orders of magnitude on both sides of the equation. Some results from the application of this model may be

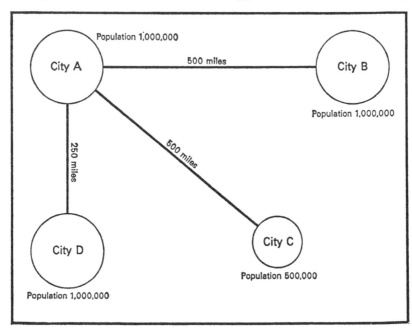

7.3 *Gravity model map. (Source: Taaffe and King 1966.)*

illustrated by the hypothetical example shown in fig. 7.3. The model would predict that the highest volume of interaction in the system would flow between cities A and D ($10^6 \times 10^6/250$). The second largest volume would flow between A and B and would be half the volume of that between A and D ($10^6 \times 10^6/500$), and twice the flow between A and C ($10^6 \times 5 \times 10^5/500$).

The most frequent measure of mass is population size but this may not be appropriate for all studies of movement. Studies of commodity movements may measure destination by population size (demand) and origin by the surplus production (supply), but we have already seen that in studies of shopping behaviour the mass of destination is sometimes more appropriately measured by an index of shopping centre attractiveness. The choice of measure depends ideally upon the problems to be studied but practically upon the data available. A similar set of considerations surrounds the measurement of distance which may be interpreted in many ways, including the actual or perceived geodesic

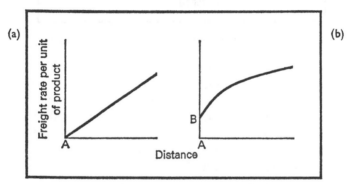

(a)　　　　　　　　　　　　　　　　　　　　　　　(b)

7.4 *Distance and freight rates.*

distance, time distance and economic distance or the cost of transport. But each of these interpretations may be modified by political distance. Movement across political boundaries normally entails an increase in all forms of distance and direct government interference with trade flows and so can cause substantial modifications to so-called 'free-trade' patterns of interaction.

Freight rates provide a measure of the cost of overcoming distance. At their simplest, such freight rates are proportional to distance (fig. 7.4a); but because they involve handling costs at either end of a journey, together with the costs of investments in handling facilities, the relationship of freight rates to distance is not simple. In fig. 7.4b AB represents the handling costs, but as the distance increases the terminal costs become a smaller and smaller proportion of the total and so the curve tends to flatten out. Furthermore, different modes of transport offer different distance/freight rate relationships and so tend to provide economically competitive services over specific spatial scales (fig. 7.5). However, the construction of continuously variable freight rates is both administratively costly and difficult for the consumer to understand. As a result, stepped freight rates, in which certain freight rate zones are

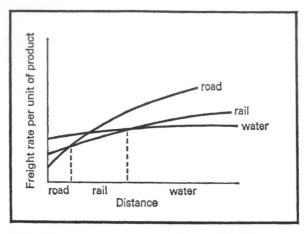

7.5 *Distance, transport modes and freight rates.*

grouped together (fig. 7.6), are more common. Commodity classifications also affect freight rates by generating different levels of cost. Thus a small, fragile, high-value, irregularly consigned commodity is likely to cost more to move per weight kilometre than a large, bulky, low-value, regular consignment because of differences in service necessitated by the two transactions. Furthermore, high-value commodities have a much lower price elasticity of demand for movement than do low-value

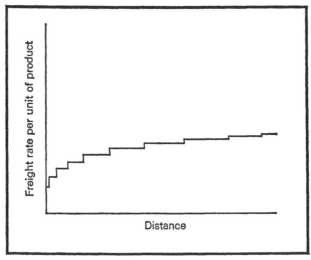

7.6 *Stepped freight rates.*

commodities. This is because the cost of transport is a much lower proportion of the total cost of the high-value commodity and so tends to be less sensitive to changes in freight rates. Clearly distance and transport cost are not related together in any simple fashion.

The empirical application of gravity models has demonstrated that movement tends to decline with distance raised to a power, rather than with distance weighted or multiplied by a constant. Thus a modified gravity model may be expressed as:

$$I_{ij} = K\frac{M_i M_j}{D_{ij}^n}$$

where n = an exponent

This distance exponent is normally derived by empirical means and varies from place to place and from commodity to commodity. Häger-strand (1957) has pointed out that differences in its value may relate to real differences in the transport network between areas which can affect the ease of overcoming distance. Thus European movement gradients may be expected to be steeper than their American counterparts (Haggett 1965).

An example of regional variations in distance friction as measured by the size of the distance exponent for originating road traffic in Britain is shown in fig. 7.7 (O'Sullivan 1970). This map suggests a certain urban–rural dichotomy with high values in rural areas and low values in urban areas. Related to this finding, peripheral areas of the economy have high values and central areas low ones. A coefficient of determination of 0·48 was found between the independent variable of population–miles and the spatial variations of the distance exponent, showing that the movement patterns in the periphery do differ significantly from those in the centre (Chisholm and O'Sullivan 1973). The size of the distance exponent indicates the rate at which interaction would decline if all other conditions were equal; the high values in the peripheral areas indicate that they attempt a greater degree of self-sufficiency or closure in order to minimize the costs of long trips to the centre. However, the spatial variations in commodity composition also affect the value of the exponent. This tends to increase for low-value, bulky commodities and decrease for high-value, compact commodities. A gravity analysis of air passenger traffic among the 100 largest centres in the U.S.A. (Smith, quoted by Taaffe and Gauthier 1973) produces similar conclusions and reveals a spatial concentration of low-distance exponents in the Manu-facturing Belt. Air traffic to other cities from New York and Chicago declines only with the square root of the distance, whereas from western cities like Denver, Seattle, Houston and Salt Lake City, traffic declines more rapidly than the square of the distance.

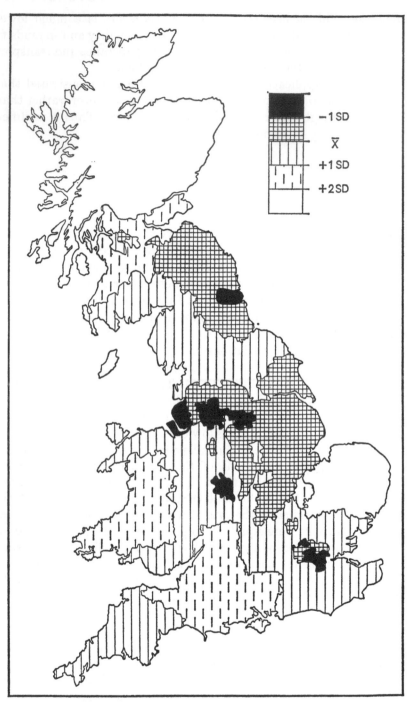

7.7 *Distance friction in Britain. (Source: based on O'Sullivan 1970.)*

Exponents might also be appropriately applied to measures of mass. For example, the mass of a place is normally considered to be positively related to interaction although in some cases agglomeration diseconomies may operate to repel movement. In this case a negative exponent could be applied to the mass variable and to the distance variable in order to account for the frictional effects of congestion upon movement. Yet many gravity models remain essentially empirical constructs: they describe and predict a pattern of spatial interaction, but seldom explain why this pattern develops. 'It is probable that the theory behind the gravity concept could be explained in terms of optimizing behaviour, such as attempts by individuals (or society) to minimize cost (or effort) or to maximize utility (or satisfaction). But few gravity analysts have tried to probe the existence of such a link by exploring the nature of the theoretical base . . . . of their models' (Richardson 1969, 135).

Intervening opportunity models
The concept of intervening opportunity was introduced by Stouffer (1940) in a study of intra-urban migration in Cleveland, Ohio. He confirmed the hypothesis that the migration of people from a zone in the centre of the city to a peripheral zone is directly related to the opportunities available there and inversely related to the number of opportunities intervening between source and potential destination. Stouffer argued that the use of distance as an explanatory variable obscured the important relationship which exists between trip distribution and the spatial incidence of opportunities. Distance only operates in an indirect way because the number of opportunities is normally some positive function of distance. This concept has been used in opportunities models of trip distribution in urban areas. The general form of these models is given by:

$$I_{ij} = G_i P_j$$

where I_{ij} = interaction between zones i and j
G_i = number of trips originating in zone i
P_j = probability of a trip stopping in zone j

The probability variable is a function of the number of intervening opportunities between areas i and j.

A study of the journey to work from London boroughs (Clark and Peters 1965) found that, for 1951 data at least, the effect of distance upon travel patterns was marginal by comparison with the distribution of opportunities between origin and destination, despite the fact that the commuting population of the peripheral boroughs had to travel further to reach a given number of opportunities. The study concluded that the spatially variable opportunities-distance relationship produced only slight variations in the opportunities–movement relationship for commuting distances of up to twelve miles.

Regression analyses

The design of regression analyses may be based upon notions derived from gravity and intervening opportunities models, whilst the explanatory power of these models may be statistically tested by placing their variables within a regression equation which also allows the inclusion of many other variables.

The building of a regression equation to explain the international trade of Italy has been described by Yeates (1968), but Tinbergen (1962) and Linnemann (1966) have attempted to extend the analysis to a world scale. The latter, and more comprehensive, study analysed eighty countries representative of a broad spectrum of levels of economic development. The flow of trade (T_{ij}) between any two countries may be expressed as follows (34):

$$T_{ij} = f \frac{(E_i)^n (M_j)^n}{(R_{ij})^n}$$

where E_i = total potential supply
M_j = total potential demand see p. 115
R_{ij} = resistance to the flow

Clearly this is very similar to the gravity model but the variables were operationalized in a multiple regression equation taking the following general form (36):

$$T_{ij} = f \left(\text{G.N.P.}_i^n, \ P_i^{-n}, \ \text{G.N.P.}_j^n, \ P_j^{-n}, \ D_{ij}^{-n}, \ F_{ij}^n \right)$$

where G.N.P.$_i$ and G.N.P.$_j$ = Gross national products of countries i and j
P_i and P_j = Population of countries i and j
D_{ij} = Geodesic distance between i and j
F_{ij} = Preferential trade factors between i and j
n = an exponent

Another variable measuring the commodity composition of trade was also included in a form appropriate for measuring the specific complementarity between an exporter's composition of exports and an importer's composition of imports. Most of the multiple regression equations using different parameter estimates were found to be very successful in explaining flows and achieved multiple correlation coefficients averaging about 0·80, although the addition of the commodity composition variable and more refined preference variables produced only marginal improvements in their explanatory powers.

Linear programming

By contrast with the other techniques of analysing trip allocation between origins and destinations, linear programming solutions to the transportation problem compute an optimal pattern of flows with which the actual flow pattern can be compared, evaluated and predicted. The extent to which the use of an optimal solution provides an accurate prediction will depend upon the behaviour of the movement decision makers. Linear programming assumes the existence of perfectly rational economic men and so actual deviations from its solutions also provide an estimate of the economic rationality of decision-making behaviour in movement allocation.

Given quantitative information about the spatial distribution of demands and supplies and the unit cost of transport, it is possible to compute an allocation of commodity flows (matching points of demand and supply in a set of dyads) which minimizes the total cost of transport. The operational procedures for undertaking this exercise have been outlined elsewhere (Scott 1971) whilst Taaffe and Gauthier (1973) provide a clear description and a number of empirical applications of the transportation problem. We shall restrict our attention here to some of the economic features characteristic of the simple transport problem. It is a relatively easy process to allocate commodities from supply to demand points such that all demands are satisfied, but optimal solutions require the minimization of transport costs. Under conditions of perfect competition the difference between prices at the origins and at destinations must be equal to the unit transport cost (fig. 3.5, p. 39). Any firm attempting to make super-normal profits on the transaction would be pushed out of the market by the entry of competitors. Wherever the difference in price between either end of a dyad is less than the transport costs, losses would be incurred by allocating commodities to the dyad; the opportunity cost is low and negative and so there is no incentive to transfer commodities from an alternative allocation. Conversely, if the price differential is greater than the transport costs, profits could be made by shipping commodities: opportunity costs are high and positive. An optimal solution is one in which all opportunity costs are negative and the profit of the marginal producer engaging in trade is zero.

Linear programming assumes conditions of perfect competition or pure monopoly, thereby ensuring complete centralized control and a homogenous product. These assumptions are far removed from reality but Scott (1971) suggests that only those systems of movement, like international trade, which are characterized by special interest groups, are inappropriate subjects for analysis. In fact a study of interregional trade in Britain (Chisholm and O'Sullivan 1973) concluded that prediction by linear programming was more accurate than prediction by gravity models.

THE STRUCTURE OF COMMODITY FLOWS

The models described above are concerned with the attenuating effect of distance upon movement and with their allocation between origins and destinations. But they treat these trade flows as though they all pass automatically through some remote controlled, central clearing market in their journey from sellers to buyers. In fact, the spatial structure of trade flows is far more complex and may be generalized by a commodity flow data-matrix (Smith 1970).

Such a matrix contains rows as origins and columns as destinations; the origins and destinations can be more or less spatially aggregated depending upon the aim of the study. The extreme form of aggregation is exemplified by a regional or national input–output matrix in which all origins and destinations are grouped together and the interest switches to inter-sectoral flows. Interregional input–output tables provide more spatial detail but these in turn are aggregates of a large number of point locations (towns, ports, route junctions) acting as origins and destinations. An alternative modification of the commodity-flow matrix involves the reduction in the number of *either* origins *or* destinations. The case of one origin and many destinations is exemplified by the study of port forelands and that of one destination and many origins by the study of the hinterland of an outdoor recreation resource.

As suggested above, the scale of the flow analysis does not affect the

TABLE 7.7 *Components of commodity flows*

Scale of Flows or connections	Volume	Terminals or Nodes
international flow	A	exporting and importing nation states
interregional flow	B	producing and consuming areas
internodal flow	C	urban centres or other transport terminals
internodal links	D	shippers or traders at terminals
internodal strands	E	consignments between traders (shipments and receipts)

Source: Based on Smith (1970, 405)

conceptual framework for study. However, important scale distinctions must be made for the detailed analysis of the volume and composition of flows, the nature of their origins and destinations, and the routes that they take between origins and destinations, at different scales of analysis.

In table 7.7 the scale differential is introduced as a function of the type of flows rather than of spatial scale differences (cf. table 7.2, p. 106). In international and interregional studies, origins and destinations, volume and composition of shipments are known, but accurate distance and route characteristics are not, whereas in internodal flows, routes and distances are known but detailed volume and value data on the composition of flows are more difficult to obtain. At more detailed scales of analysis – internodal links and strands – difficulties of confidentiality relating to individual firms restrict the availability of both types of information. However table 7.7 suggests that the internodal flows are made up of the sum of internodal links which in turn consist of the total of individual consignments from one trader to another such that:

$$\Sigma E = D$$

and $\qquad \Sigma D = C$

and $\qquad \Sigma C = B$

but $\qquad \Sigma B \neq A$

because much of B will be consumed within the home market although it may be expected that A is at least partly a function of B.

Within this framework, commodity flow analysis is concerned with three central issues (Smith 1970): the diagnosis of differences between predicted and actual flow volumes over individual routes; the efficiency of a given flow pattern; the spatial structure of commodity flow patterns. The first of these topics has been extensively investigated by the use of gravity and regression analyses, looking in particular at the residuals from predicted flows. Efficiency in flow patterns is effectively measured by linear programming, whilst the structure of commodity flows is concerned with the description of the spatial pattern of connections within an economy. This latter concept draws attention once more to the two-way interactions between movement and spatial structure.

In North America, despite the fact that the major lines of physical relief run north–south, the socio-economic landscape is oriented in an east–west direction. The major productive force of the American industrial belt stretching from Illinois to New England (containing about 60% of U.S. manufacturing industry and 40% of the population) generates the largest amount of internal movement and aligns the major routes of the rest of the country. Bulk commodities like coal, iron ore and petroleum provide the major interregional element in rail freight tonnage and their sources are the major origins of traffic attracted by the core area of the manufacturing belt. Other zones of traffic generation include the principal agricultural regions, forest areas and fruit and early vegetable production regions (Ullman 1957).

A comparable study of commodity flows (Berry 1966) attempted to isolate the underlying principles determining the pattern of connection within the Indian economy. The maps of commodity flows showed that they were dominantly intra-regional and oriented towards the major metropolitan centres. This suggests a division of the sub-continent into a set of regional economies focussed upon these centres. In addition, there was an inter-metropolitan movement of commodities emanating from the modern manufacturing sectors of the economy, and movement into the metropolitan regions from external origins based upon zones of agricultural specialization or resource-based industrial complexes. The metropolitan regions act as origins, destinations and reorganizers of flows, and the efficiency with which they perform this task is of direct importance to the fortunes of the Indian national economy.

This role of urban nodes as important organizers of commodity flows has also been demonstrated in a study of interregional trade in Nigeria (Hay and Smith 1970). A disproportionately large quantity of trade passes through the major towns, partly because they are the major spatial concentration of elements in the Nigerian economy and hence the major centres of demand and supply, and partly because it is only in these nodes that adequate fiscal and physical facilities exist to handle the flows.

Conclusions

In this chapter emphasis has been placed upon the links between the operation of the economy and the demand for movement. Movement and locational specialization are intimately related and the ability to move both within and between areas is a fundamental determinant of the nature of the local economy. We have already seen that inter-personal variations in the ability to move result in socially unjust inequalities. The same is true of differences between space-economies, especially at the international level; even where movement is possible, the gains from trade are not shared equally.

Some of the variables which control the generation and allocation of movement have been examined in this chapter along with some simple models used in attempts to examine the effectiveness of these variables. But the chapter has allowed itself the luxury of abstracting from the real world in so far as it has not taken the spatial or organizational structure of the movement agencies into account as powerful constraints upon the flows generated. This situation is modified in the following chapters where the spatial structures of market facilities and transport networks are considered in more detail.

8 Market centres

Two sets of markets operate within economies: the market for goods
and services and the market for factors of production. Both sets link
buyers and sellers together and so facilitate the exchange of commodities passing between them. At its most complete an 'efficient
marketing system will provide a means of determining the needs of
consumers (buyers) – the types of commodity they require, the time
and place at which they require those commodities and the prices they
are prepared to pay; it will transmit this information about requirements promptly and accurately to the producers (sellers) through the
mechanism of price quotations (offers and bids); and it will arrange for
the transfer of commodities from seller to buyer in the required quantities and at the lowest possible cost' (Britton *et al.* 1969, 11).

These functions of the marketing process imply a considerable degree
of coordination between buyers and sellers. Chapter 3 examined the
way in which the price mechanism and its administered alternative
attempt to achieve this coordination. But the successful operation of
either method within a space-economy depends upon the existence of
a network of market places in which buyers and sellers can meet and
through which goods and the payment for them can be transmitted.
The demands of the marketing process are largely responsible for the
spatial and functional characteristics of the network of market places,
but this in turn is a major influence upon the efficiency of market
operations. It is this general two-way relationship with which we are
primarily concerned in the present chapter (see also Dawson, 1973).

Specific attention, however, is limited to a consideration of the
market places for goods and services only – a limitation which reflects
the scarcity of geographical analyses of the markets for factors of production. It is a strange paradox that whilst economic geographers have
given distinguished attention to the study of resources and factors of
production they have until recently largely neglected their associated
market places. Conversely studies of the markets for goods and services
abound whilst certain aspects of the geography of consumer behaviour
remain undeveloped (see chapter 4). Manners (1971) has drawn attention to the evolving spatial structure of the world market for iron ore
(which operates without any recognized market places), Thoman and
Conklin (1964) have noted the influence of the physical handling and

control of commodity flows upon the spatial structure of international trade and Jeans (1967) has examined the location of commodity markets in London.

Although the study of the market places for goods and services has been an accepted part of the subject matter of economic geography for some time now (Applebaum 1954), it was only during the nineteen-sixties that it acquired any significant place in the literature (Berry 1967; Vance 1970). In this chapter we present, first, a discussion of four key concepts associated with market centres in general and then move on to consider briefly the geography of retail and wholesale market centres. The three sections are introduced separately but they are nevertheless linked together by a hypothesis of sequential development.

Market centres

In dealing with market centres we are concerned with the actual places in which the market for goods and services is organized. A market place may be defined as a public gathering of buyers and sellers meeting at appointed locations and at regular times. Of all the many aspects of the study of market centres there is space here to mention only four.

CENTRAL-PLACE THEORY AND MARKET NETWORKS

The function of a central place is to provide a wide range of goods and services to the population surrounding it, but the existence of a central place is dependent upon its access to a minimum or threshold level of demand able to support the service of distribution offered by the centre. We have already seen how, under certain simplifying assumptions, a market area may be delimited around a central market as increases in distance have the effect of increasing price and decreasing demand (fig. 3.3, p. 37). The limit of the market area marks the maximum extent, or range of demand, accessible to the market centre, and this range must be equal to or greater than the threshold value of the good or goods distributed from the market place if it is to remain economically viable.

Under conditions of perfect competition any unsatisfied, effective demand beyond the range of the established market will generate more markets. They will continue to be established as long as their threshold exceeds their range. As the number of markets increases, market areas will contract in size up to their threshold limits and change in shape from circles to hexagons, so that all parts of the area are served but no overlapping of market areas occurs. The end result, on an isomorphic plain with an even distribution of population, is a regular network of equally-spaced market centres, each surrounded by an equal-sized hexagonal market area.

Not all goods have the same threshold value. Some require a much larger minimum level of demand than others and so it is impossible to provide all goods at all market centres. Only the lowest-order goods with very small thresholds can be distributed from every market. Markets dealing in higher order goods are found much less frequently and Christaller (1966) suggested that market centres are arranged in an orderly hierarchy in which a centre of a given rank supplies the goods appropriate to that rank and all lower-order goods.

The number of markets offering goods of the highest order ('A' ranking centres) will depend upon the total volume of demand for these goods within an area. The 'A' rank centres will also be the sole sources of supply for any other goods whose range is too large for the establishment of more frequently distributed outlets. As the rank order of goods is descended their thresholds and ranges diminish. The highest ranked good offered by the next rank of market centres ('B' rank centres) will be hierarchically marginal because its range will be small enough to permit its supply from the more frequently occurring 'B' centres.

This allocation of a ranked set of goods to a spatial set of market places continues until all effectively demanded goods have been provided with outlets from the appropriate rank of centre. The spatial system of market centres is then maintained by the energy of effective demand at the market and the ability of firms to supply the market with the goods desired.

ORIGINS OF MARKET CENTRES

Even the most cursory acquaintance with market studies reveals the need to examine the various ideas or theories about market origins, although much of the existing work on this is still very tentative. The most common, orthodox theory starts from the individual's propensity to barter; deduces from this the necessity for local exchange, the division of labour and local markets; and infers finally the necessity for long-distance or at least external exchange or trade. An alternative theory reverses this sequence entirely, stating that markets can never arise out of the demands of purely individual or local exchange, for markets depend on trade, which means external trade. Markets then, are not the starting point but rather the result of long-distance trading, itself the result of the division of labour and the variable geographical location of goods. Thus for the traditional, still largely peasant economies of contemporary sub-Saharan Africa, it has been suggested that the origins of traditional or indigenous markets are explicable only in terms of a hypothesis which views markets as the end product of long-distance trade and contacts, together with the conditioning factors of the kind of physical security that can ensure the market peace and a sufficiently high population density to make regular face-to-face contacts of people practicable. In Yorubaland (West Africa) the location of markets, their commodity structure and their location in relation to the pattern and hierarchy of settlements – all suggest that markets there are related genetically to external trading contacts (Hodder 1964). Similar conclusions have been reached in work elsewhere (Berry 1967; Jackson 1971).

This second, unorthodox, viewpoint about market origins follows the work of Polanyi *et al.* (1957) and Pirenne (1953), and is given further support by much contemporary work in this field. Such a viewpoint is relevant, not only to speculative thinking about the past and to modern theoretical discussions; it is also relevant to contemporary ideas on development planning, more particularly where these refer to the problem of inducing self-generating change in a market economy among those peoples of the developing world who as yet lie largely outside modern economic forces. For instance, if this second viewpoint about market origins is valid, it means that only where there is an effective political administration to maintain physical security, a sufficiently high density of population and long-distance lines of external contact, especially good trunk roads, are efforts to bring about the transformation of backward rural economies likely to be effective. And where these conditions are not fulfilled, then their provision must be a prerequisite for any significant economic advance.

PERIODIC AND DAILY MARKETING

Much of the present discussion on market places emphasizes the fundamental distinction between periodic and daily marketing. Periodism is an essential element of the local, especially rural, indigenous market structure in most developing countries today, as indeed it was of medieval Europe. The frequent repetition of activities in a periodic market is presumably due to lack of storage facilities, elementary transport facilities, and a population density too low to support continuous trading. Markets occur at intervals of two to eight days in different parts of the world, but unfortunately our knowledge of the distribution of these types of periodic market is scanty, and in very few areas is it possible to construct even a simple distribution map of the types of periodic markets (fig. 8.1).

It is also known, though here again data are very few, that most periodic markets operate in groups, in what can be termed ring systems or circuits, which work in such a way that neighbouring markets do not compete with each other on the same day. Moreover, such a system means that the timing of marketing activities is evened out over the whole circuit so that no settlement is far from a market for more than a day or so. This integrated pattern and timing of markets is logical and convenient and is, in some form or another, characteristic of most countries. It is clearly a wholly indigenous and logical phenomenon,

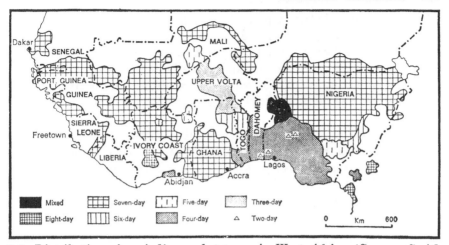

8.1 *Distribution of periodic market types in West Africa. (Source: Smith 1971.)*

expressing an intelligent mutual self-interest among neighbouring settlements, peoples or market authorities. And once again, the similarity between such market circuits in the developing world today and, for instance, in medieval East Anglia (Dickinson 1934) or Derbyshire (fig. 8.2) is most striking.

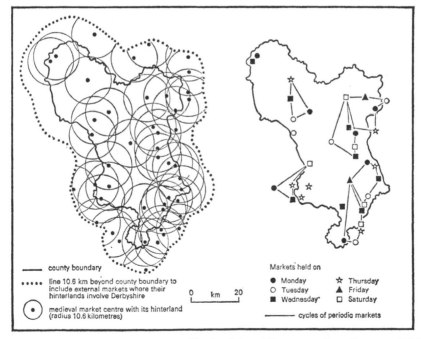

8.2 *Medieval market circuits in Derbyshire. (Source: after Coates, 1965; Fox 1970)*

The distinction between periodic and daily, continuous marketing usually involves very much more than simply the timing of marketing activities. What data are available point to differences in location and functions as well, periodic markets usually being rural and acting chiefly as collecting points for surplus agricultural and cottage industrial products, while daily markets are usually urban and concerned chiefly with retail buying and selling. This distinction can most often be seen in and around a large city, which may be surrounded by a number of periodic markets, organized into integrated systems of market circuits and operated largely as suppliers of farm produce and other locally produced commodities to the urban population. Inside the city, on the other hand, the major markets are usually large daily markets.

In purely economic terms a general explanation of market periodicity may be made within the analytical framework of the theory of the firm (chapter 5). Thus periodic marketing may represent an attempt by traders to reduce average costs, increase average revenues or both. Stine (1962) concerned himself with an analysis of market mobility as a means of economic survival (fig. 8.3). Provided that the range of a market outlet is greater than its threshold (a), a full-time and permanent location for that outlet is economically viable. But if threshold exceeds range (b, c_1, c_3) sales would be insufficient to cover costs. One response to this problem is market mobility. By rotating the outlet around a series of market places costs may be spread over a greater volume of sales. In this way threshold and range could be equated and costs covered. The number of market places in a given circuit (c_2, c_4) is a positive function of threshold and a negative function of range and of the attractiveness of the market places themselves (see pp. 58–63 for a discussion of these interrelationships).

This explanation of periodicity is partial in the extreme although it has been widely quoted and accepted (Berry, 1967). It emphasizes market mobility to the exclusion of the possibility of part-time marketing which in fact may be a more rational response if the costs of mobility are significant. Furthermore periodicity may not result only from the drive towards economic viability; the maximization of profit or the elimination of competition by lowering retail prices or raising producer prices are equally legitimate objectives. This means that Stine's case – mobile marketing for viability – is merely a special case of periodic marketing and in this sense is inadequate as a basis for the understanding of periodic marketing and its evolution over time (Hay, 1971).

Perhaps the most important problem requiring much deeper investigation here is to understand exactly how changes in the periodic daily marketing structure take place, and in particular in what way periodic marketing eventually becomes continuous trading. On the basis of the available evidence, it seems likely that the trend from periodic to daily marketing is a gradual one and commonly involves a progressive shortening of the period between market days so that, for instance, an eight-day market becomes a four-day, then a two-day and finally,

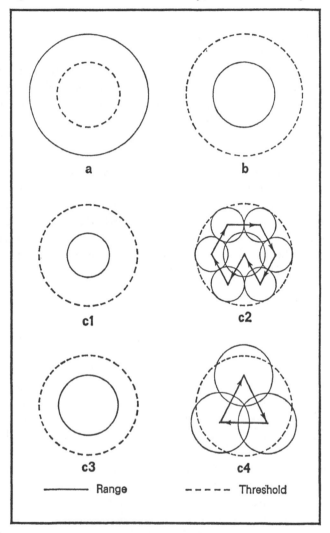

8.3 *The trader-consumer hypothesis. (Source: Stine 1962.)*

perhaps, a daily market as population density, urbanization and accessibility increase. Some periodic markets may, of course, go the other way, decreasing in frequency and perhaps dying out altogether. But the normal and logical trend in periodic marketing seems to be towards daily marketing, which is particularly associated with urban forms of settlement. It is indeed remarkable how the scanty material on this – from China (Skinner 1964), India and Africa, as well as from medieval Europe – supports the view that increasing frequency and finally the stability of marketing operations is inevitable in conditions of increasing population density, urbanization and economic development.

SEQUENTIAL DEVELOPMENT

It is significant that much of the contemporary increase of interest in markets as places comes from those social scientists who are primarily concerned with central-place studies. Certainly one of the best bibliographies of material relating to markets and fairs is to be found in a volume on central-place literature (Berry and Pred 1965). At first sight, and especially in the light of a frequent lack of correspondence between rural periodic markets and the location, size and hierarchy of settlements, it might seem that central-place models or theories have little if any relevance to market place studies today. But it is to be doubted whether such a viewpoint can now be held in the face of the remarks made above about the trend towards daily, urban marketing. One cannot ignore the growing volume of material on markets and central-place theory from China, India, Yorubaland and Iboland (Skinner 1964; Dutt 1969; Hodder and Ukwu 1969). Moreover, as Garner (1967) has pointed out, the periodic central-place is a wholly valid concept within the context of modern central-place theory.

Much of the evidence now available, in fact, suggests strongly that all types and operations of markets may be seen as intermediate stages on a single, albeit many-stranded continuum from the most elementary to the most complicated economies. In spite of the wide differences in period, in the original structure of society and economy, as well as in the time span of the processes, the similarities in market-place development throughout the world are very striking. The history of markets in Western Europe can be traced to the rise of fairs in the tenth century following the breakdown of local isolation and the expansion of commercial contacts. It can be followed through their relative decline in the face of increasing urbanization, the growth of urban commerce and, later on, the development of overseas trade. Although the various institutions – market place, fair and market town – existed side by side well into modern times, the town eventually took over most of the functions of the fair. On the other hand, the local market, where extant, has no more than a peripheral role; it becomes an occasional mart for the local disposal of certain agricultural products or exists only as part of a neighbourhood shopping district.

Viewed from the standpoint of the evolutionary history of the Western European marketing system, the economic landscapes of many developing countries today show anachronistic juxtapositions of institutional features. Change, especially following contact with European trade and control, may have been evolutionary rather than revolutionary, but it has certainly involved and must still involve the telescoping and overlapping of the processes and time span of social, economic and technical transformation. It is possible, as several writers have done, to generalize empirically about market evolution in the economies of

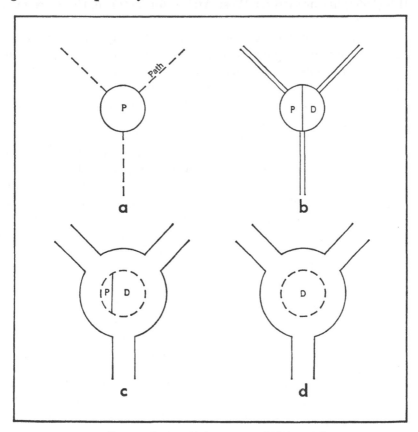

8.4 *Sequential development to daily market.*

developing countries and to devise a number of sequent stages. Fig. 8.4 (a–d) shows the shift, from periodic (P) to daily marketing (D), as accessibility improves with the change from primitive paths to higher capacity roads. Permanent retail shops, often located around the perimeter of the market place, and specialized wholesale markets are the end results of such a process which has been observed in several case studies, especially in the cities of less-developed countries (e.g. Hodder and Ukwu, 1969).

The immediate and practical importance of all these considerations for urban and rural planning and for the understanding of the operation of an economy needs little emphasis. Rural periodic markets continue to dominate the collecting and wholesale marketing activities in the rural areas of so many countries; and there is little doubt that daily markets will for long continue to dominate the retail structure of cities in the less developed world. Moreover, if what has been said above is true, then the urban daily market place is likely to increase in importance. As Hill (1963) has argued for West Africa, for instance, there are many powerful practical reasons, such as increased urbanization, improved transport, higher purchasing power, increased occupational specialization and similar developments in economic structure, for presuming that the total quantity of goods sold in the urban daily markets is increasing proportionately far more quickly than the size of the population. At the moment we know least about daily markets, so that as urbanization proceeds we know less and less about the provisioning of the people. It is perhaps too easily assumed that markets are out of place in a modern city and it is often forgotten that the change from markets to shops in most European towns took place only very slowly over a century or so. Moreover, it is by no means certain that open markets are not ideally suited to the prevailing physical, social and economic conditions in developing countries. It can also be pointed out that the market place still plays a not unimportant role in the retail and wholesale trade of Western cities, as has been demonstrated for London (Buzzacott 1972). But for the developing world, certainly, town planning can perhaps realistically be developed only on the assumption that market places will continue to dominate urban retail structures for some time to come. Yet little can be done while the basic data about markets in these areas are so rarely available. Apart from the above-mentioned lines of research requiring immediate attention, there is an urgent need for the kind of work on market catchment areas undertaken by Dutt (1965). For it cannot too often be emphasized that markets in the towns of the developing countries 'play the vital role, rather than shops and shopping centres, as media in the social and economic life of the people' (Smailes 1964).

Retail and wholesale markets
It is now time to refer briefly to the work being carried out on the two end products referred to in the above argument – retail shopping and wholesale marketing activities – with special reference to developed urban economic systems. As for the literature on retail shopping, much of this has been discussed by Berry (1967) and his synthesis is an essential starting point for any study by an economic geographer of retail shopping. As Ginsburg (1967) notes, Berry attempts to indicate principles concerning the spatial distribution and organization of marketing. In Berry's own words, his thesis is 'that the geography

of retail and service business displays regularities over space and through time, that central-place theory constitutes a deductive base upon which to understand these regularities, and that the convergence of theoretical postulates and empirical regularities provides substance to marketing geography and to certain aspects of city and regional planning' (1967, vii).

Another possible approach to the study of retail market centres has already been implied in the previous section – the study of the processes involved in the change from the simplest to the most sophisticated forms of urban retail establishments. Here, although the location, hierarchy and spatial arrangement of such phenomena are not ignored, the emphasis is rather different. This approach is, as yet, less susceptible to theoretical treatment, nevertheless it may be more fully integrated into other processes in the economy. It also takes into account the obvious fact that whereas in smaller centres successful shops and shopping centres locate themselves – the decisions leading to these results being taken mainly by dispersed individuals using arguments that they did not make explicit – in larger centres and in 'modern' situations, such decisions may be taken by planners – local authorities, retailing firms, or developers – who reach conclusions about substantial groups of shops and justify their conclusions by written arguments. Such an historical-analytical or cross-cultural comparative approach to the study of retail marketing has so far received relatively little attention from economic geographers.

As for the geography of wholesale marketing, perhaps the best statement on the work now being carried out in this field by economic geographers is to be found in the work of Vance (1970). Vance examines the factors underlying the distribution and organization of wholesaling activities. Strongly historical and analytical in its approach and reaching back well into the middle ages for an understanding of the role of wholesaling, Vance's book opens up an important and as yet very little studied aspect of economic geography.

The distinction between wholesaling and retailing is by no means always clear, more particularly in small-scale rural economies in developing countries, but a more precise definition than that normally implied – that the wholesaler connects the producer with the retailer – is demanded. Beckman and Engle (1937) proposed three basic criteria to distinguish wholesaling from retailing: (i) the status or motive of the purchaser – wholesaling involves the sales of one entrepreneur to another, intended for resale by the second; (ii) the quantity of goods involved in the transaction, wholesaling being more concerned with accumulated demand rather than unit demand; (iii) the method of operation of the concern. Also useful, perhaps, is the Standard Industrial Classification (1968) definition of wholesale trade as involving

F

'establishments or places of business primarily engaged in selling merchandise to retailers; to industrial, commercial, institutional, or professional users; or to other wholesalers; or acting as agents in buying
merchandise to such persons or companies'.

The notion relating wholesaling sequentially to the trend from
periodic marketing (with its dominantly wholesale characteristics) to
daily marketing (with its dominantly retail functions) – wholesale
markets eventually being separated out from urban daily marketing –
has already been presented. Superficially, at least, the relevance of
central place theory to the geography of wholesaling seems to bear
careful study. But, after carefully considering central-place theory,
Vance (1970) concluded that the theory failed to provide 'the system
of analysis needed to account for the general structure and location of
wholesale trade . . . in particular because it fails to handle trade prac-
tised by agents' (71). While it is true, as Berry's (1967) work demonstrates, that the assumptions of central-place theory allow some judgment of the total volume of trade expected at a particular spot and the
geographical reach of that centre's appeal, none of these measures
applies to wholesaling. Trade dealings in wholesale trade tend to be
carried on in a most confidential way, and there is no real upper range
of distance in the sale of goods. Furthermore, confusion is created
through the abstract quality of many transactions. The whole of
experience, indeed, suggests that we should use a historical rather than
a temporally abstract approach, and substitute induction for the
rigorous deduction of central-place theory.

Vance suggests that the paradigm for wholesale trade is not the
central-place model but what he calls the mercantile model, the important distinction between them focussing on the source of growth and
change. Whereas endogenic forces are dominant in central-place
analyses – growth coming largely through enhanced demand on the part
of local consumers – in the mercantile model the dominant forces are
exogenic, originating from a wide range of sources.

Conclusions
It is probably safe to assume that the participation of economic
geographers in the field of marketing geography will increase rapidly
over the next decade or so. In both simple and complex economies the
relevance of such work is now generally recognized. As Applebaum
(1954) initially stated the case, the well-developed topical fields of
economic geography in the early nineteen-fifties were at that time,
with the exception of transportation geography, concerned with the
production of material goods. But goods must not only be transported
from production to consumption areas; they must also be transferred
from the hands of producers, by collection and subsequent distribution,

into the hands of consumers. Marketing studies provide, indeed, a suitable focus for the 'geographies of consumption, production and cities, and links them . . . through a theoretical system' (Ginsburg, in Berry, 1967, vi).

9 Transport

Transport networks are the physical expressions of integration within and between economies, but they are both cause and effect of the movement demands induced by this integration. The purpose of this chapter is to consider some aspects of the two-way relationship between transport demand and supply, and to examine the spatial characteristics of transport networks as the expression of functional integration within space economies.

Demand and supply

The generation and distribution of spatial interaction together represent the demand for transport. The supply response to this demand is the creation of transport networks and the assignment of movements to them. But the demand for transport is not an independent variable affecting the supply of transport because both the generation and the distribution of movements are closely affected by the structure and efficiency of transport networks.

The prediction and evaluation stages of movement analysis (p. 108) match movement demands with transport supply, and future demands with the resources available to increase that supply. If transport could be provided under conditions of perfect competition, an optimum network would automatically result by equating the marginal cost of the last unit supplied with its marginal revenue. Obsolete sections of the network would be withdrawn where $MC > MR$ and new sections would be supplied, with the appropriate modal characteristics, wherever the desire lines of movement between places revealed that normal or super-normal profits could be made by supplying a service. Although the early development of railways in Britain was characterized by a highly competitive market, with many buyers and sellers, the nature of the product prevented the operation of a perfectly competitive market.

Transport networks are both 'lumpy' (because they necessitate the production of a high minimum quantity before any unit of output is economically viable) and capital intensive (because of their substantial engineering input which is, in part, a function of the physiography of the area through which they are constructed). These characteristics are particularly problematic in the capital-scarce less-developed countries where the proportion of public expenditure devoted to transportation investment may range from 20 to 40% (Gauthier 1970). Under these conditions rational decision-making processes concerning investment in transport facilities are vital.

Gauthier (1970) has drawn attention to the relevance of the concept of balanced and unbalanced growth (Hirschman 1958) for any understanding of the relationship between supply and demand in transport. A certain minimum level of social overhead capital (S.O.C.), which includes transport facilities, is required as a prerequisite for direct productive activity (D.P.A.), but the relationship between the two is neither precise nor technologically determined and may be balanced or

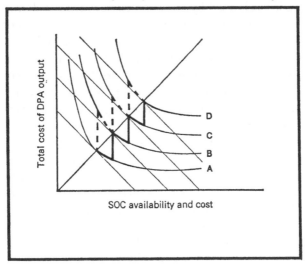

9.1 *Capital investment and unbalanced economic development. (Source: Hirschmann 1958 and Gauthier 1970.)*

unbalanced through time. This problem of balance is illustrated by fig. 9.1, in which the vertical axis measures the total cost of D.P.A. and the horizontal axis measures S.O.C. The curves A to D represent successively higher amounts of output and the aim is to generate the maximum D.P.A. at minimum resource cost in both D.P.A. and S.O.C. An optimum ratio of D.P.A. and S.O.C. is defined as the point at which an iso-cost line is tangential to the highest achievable level of output. But the lumpiness of capital investment in transport and, at least in less-developed countries, the scarcity of capital, together prevent the achievement of this balance as output expands. Two sequences of unbalanced development are suggested by fig. 9.1: one in which economic growth is stimulated by an excess of S.O.C. which may reduce the resource-cost of D.P.A. (shown by the heavy line), and a second in which growth in S.O.C. is stimulated by the pressure of demand from D.P.A. (shown in the diagram by the dashed line).

Transport is also characterized by heterogeneity. At the most basic level any two units of transport are bound to be differentiated by their location and route, but such differences are intensified by modal distinctions. Each mode of transport can offer a distinctive service, as exemplified by the differences in their cost-distance relationships (pp. 125–126). The effect of these differences may be demonstrated by the greater importance of pipelines and inland waterways over the longer distances of mainland Europe by comparison with a road-dominated United Kingdom space-economy in which average lengths of haul are much shorter. Despite this distance-induced complementary relationship between the various modes of transport, they are also strongly competitive as is evidenced by the successive waves of dominant modes of transport in the history of many advanced capitalist economies. A typical sequence of dominance has been non-motorized road transport, canals, railways and motorized road transport. In the less-developed countries railways, for long the major stimulus to economic development and symbol of colonial economic power over many areas, are gradually being replaced and extended by less expensive and more flexible road networks.

This sequential dominance is a result of the two-way relationship between transport supply and demand. For example, the increasing use of motor-cars allows their owners greater choice in residence and work place and so helps to create spatially dispersed cities which in turn demand the use of automobiles. Transport may constrain the spatial structure of an economy, but the shape of this structure is in turn a major influence upon the modal demand for transport.

Cross-sectional analyses of modal choice by consumers of transport tend to be conducted at the micro-scale of the individual decision maker and emphasize, within a variety of analytical research designs the consumer's assessment of the various attributes of the different modes (e.g. Bayliss and Edwards 1970). Spatial analyses of modal choice are rare but include studies of the modal demand for transport in the U.S.A. (Perle 1964) and Britain (Chisholm and O'Sullivan 1973).

Perle (1964) began his analysis with the premise that within the United States there are numerous transport markets which may be defined by the major commodity groups and by the economically specialized regions within the national space-economy, but that these markets are all subsets of a larger national market. The major aim of his study was to examine the relationships, in each of these markets, between the demand for transport and price as the primary demand determinant. Several levels of analysis (macro – or national demand for all commodities; meso – either national demand by major commodity groups or regional demand for all commodities; micro – or individual region commodity combinations) were conducted using data for the period 1956–60.

The results of this analysis showed a marked shift away from rail to road but a considerable variation in the extent of this shift between individual commodities and regions. For example, the price elasticity of substitution between the two modes was shown to be regionally sensitive. This measure of elasticity may be expressed as: $Q_m/Q_r = f(P_m/P_r)$, where Q_m, Q_r = quantities of goods carried by road and rail transport respectively and P_m, P_r = prices of road and rail transport.

The ratio of the quantity of goods carried by road to the quantity carried by rail is thus a function of the ratio of their respective prices. Perle found large growth rates for the motor/rail consumption ratio, indicating a commodity capture by road transport in the Southern, Middle Atlantic, North-western and South-western regions, but in the New England, Mid-western, Rocky Mountains and Pacific regions low growth rates and high elasticities were highly correlated, demonstrating that small price shifts produced marked consumption shifts and that effective intermodal competition existed in which each mode had to refrain from pricing itself out of the market.

Although Perle clearly succeeded in demonstrating the existence of wide regional variations, the causes of these differences were left unexplored. By contrast, Chisholm and O'Sullivan (1973) suggested that if rail traffic has a competitive advantage over road traffic for long hauls, then it might be expected that rail traffic would be relatively more important for the movement of commodities to and from spatially peripheral zones of the economy than for movement to and from central zones. This expectation was tested by expressing road-freight tonnage as a percentage of the total for originating and terminating traffic respectively. These percentages were then treated as dependent variables and regressed on both population miles and the mean haul for road traffic for each zone. The results showed conclusively that the general location of a zone within the British economy has no significant influence upon modal choice. But rail freight consists of a limited number of commodities, with specific origins and destinations, which suggests that the proportion of originating and terminating rail-freight traffic does vary spatially according to the location of quite specific activities.

A further difference between any good, offered in a perfectly competitive market, and transport is the latter's abundance of externalities. Unless the supply of transport is quite private and closed to outside users its economic costs and benefits flow well beyond the owners or builders of the network. These externalities may be exemplified in a spatial context by the external costs of noise and visual intrusion that a major transport facility (e.g. an urban motorway or an international airport) may generate within its vicinity. But they are much more complex than this. Wilson (1966) suggests that all investment outlets might usefully be construed in terms of their relative influence on attitudes, *vis-à-vis* their impact on labour productivity induced by an increase in the capital labour ratio:

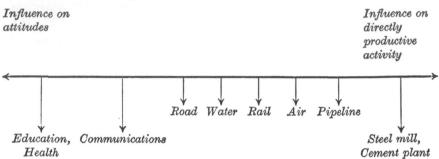

Influence on attitudes

Influence on directly productive activity

Road   Water   Rail   Air   Pipeline

Education,   Communications
Health

Steel mill,
Cement plant

The design and ownership of road networks is such that there are normally few barriers preventing the use of roads as they pass through an area, whereas pipelines rarely stimulate local use as they are normally completely closed by design and private in ownership. Furthermore, the fact that there are few economies of scale in the road transport and repairing trade also reduces the barriers to entry and so provides an opportunity for the growth of entrepreneurship.

Investment in transport also generates spatial externalities, and transport supply affects movement demands by altering the relative locations of places and by changing the connectivity of economies (see pp. 156–160). Transport-induced reductions in the time and money costs of movement reduce the influence of distance upon the spatial distribution of economic activity and upon the interaction within and between economies. They fuse market areas, reduce spatial restrictions upon the location of production and bring additional buyers and sellers into contact with one another. As a result, the locational specialization of economic activity is based more upon comparative advantage than upon market accessibility (Lloyd and Dicken 1972, 89) and so intensifies the differentiation of economic space. However, agglomeration economies intensify the comparative advantage of urban locations for the location of economic activity and so tend to override tendencies towards a more dispersed pattern of location, based upon low-cost transport, in favour of spatially concentrated patterns of production which can also take advantage of local specialization and external agglomeration economies. At first, spatial specialization of this type further increases the demand for movement and transport as intra-spatial self sufficiency is replaced by inter-spatial dependence. This phase may gradually be replaced as the large urban agglomerations increase in size and so provide an economically viable market for an ever-widening number of products. This in turn may promote internal self-sufficiency and a reduction in inter-urban transport except for the movement of highly specialized commodities, ideas and people (fig. 1.3, p. 11).

As a result of these externalities, investment in transport performs a dual function (Lachene 1965). Transport provision serves both the short-term function of satisfying the demand for movement between areas and the long-term function of helping to shape the growth of places by inducing changes in comparative advantage as a result of changes in accessibility and relative location. Such long-term feedbacks (Haggett and Chorley 1969) of transport investments may be clearly identified, monitored and evaluated at the local level. The classic study of the effect of a by-pass around Marysville, Washington (Garrison et al. 1959) showed that whilst the fall in through traffic increased the town's attraction for low-order goods and residential functions, the by-pass extended the market area for higher-order goods in the neighbouring, non by-passed, town of Everett. Transport impact studies (see also Kellett 1969) are much more difficult to undertake at regional (e.g. Taaffe 1960; 1962) and national scales (e.g. O'Sullivan 1969; O'Connor 1965 (a and b); Hall 1969), if only because of the increase in the number of variables that can intervene to complicate the relationship between transport investments and spatial adjustments. This complexity is well illustrated by the confusion surrounding the proposed investment in a major seaport facility at Portbury in South-West England (Tanner and Williams 1968).

One consequence of this dual function of transport is that there is, as yet, no consensus on the role of transportation in the development of the space economy (Berry 1959). Three possible relationships may be defined (Gauthier 1970). Investment in transport may have: (i) a positive effect upon the development process in which growth is a direct result of the provision of improved transportation facilities; (ii) a permissive effect because transport cannot, by itself, induce development but may offer the possibility of economic expansion; (iii) a negative effect in that scarce resources with a very high opportunity cost are used in producing redundant networks and so hold back expansion elsewhere in the economy. Given the paucity of studies relating movement demand to aggregate measures of economic activity, debates on the role of transport in economic development, exemplified by the classic discussion on the role of railroads in the economic growth of the United States during the latter half of the nineteenth century (Rostow 1960; Fogel 1964), will continue (e.g. Hoyle 1973). The fact that the 'lead' and 'lag' relationships of transport to economic growth can be clearly identified (Wilson et al. 1966) but rarely explained (Gauthier 1968) underlines the complexity of the relationships between transport and spatial economic development and the difficulties of supply decision making (e.g. Starkie 1970). Errors in the allocation of resources are especially likely in the transport sector, both for the reasons outlined above and because the uncertainty which surrounds transport investment provides an element of protection to political decision makers and so increases the attraction of transport investments.

Centralized transport decision-making units may make use of sophisticated techniques of evaluation like Cost-Benefit Analysis (C.B.A.) to try to take account of all the externalities associated with a major transport investment. The aim of cost benefit studies, exemplified most comprehensively in the transport sector by the study undertaken for the Commission on the Third London Airport (1971), is to take account of all the gains and losses generated by the project, including both its internal, short-term and private effects and its external, long-term and public implications.

Most criticisms of C.B.A. derive from its need to measure subjective and relative concepts like noise and to value such things as environmental or aesthetic attraction. In the former case a given, objectively defined, increase in the noise level will induce a wide range of subjective human responses related to the proportionate size of the marginal addition to noise for the individuals concerned. In the latter case there is no clearly defined market for environmental beauty so that any valuation placed upon it must be arbitrary and will tend to reflect the social norms of the valuers. Furthermore the longevity of both the construction period and the finished product, together with the size of the resource commitment, serve to intensify the difficulties of decision making associated with transport supply.

Geographical studies of the relationship between demand and supply in transport are mainly concerned with the evolving spatial form of the transport network rather than with the processes of decision making underlying the spatial structure. However, a geographical comparison of two contemporary rail nets – one developed by a state owned enterprise in South Australia, the other built by several private companies in the physically similar Columbia Basin in the United States – showed that the former was much more wasteful in the provision of routes and use of scarce resources than the latter, which was able to benefit from central coordination and control (Meinig 1962). The application of network analysis (pp. 156–165) to transport supply decision making has been exemplified by Burton (1963) in a study of the expansion of the highway network of north-eastern Ontario. Complex problems like the evaluation of changes in the accessibility of the total network and of the individual places that it connects, as a result of its extension, can be approached relatively simply by using techniques of network analysis.

By contrast to this measured form of transport expansion, the competitive chaos of early canal and railway developments in Britain (e.g. Appleton 1962) was a response to the pent-up demand for movement and the chance to earn heavy profits during the Industrial Revolution. As a result many lines were duplicated and resources wasted. Eventually, indeed, transport firms were forced to amalgamate, modify their competitive relationships and so reduce the network (Patmore 1966; Appleton 1965b). The opportunity for the making of such large and rapid profits from the production of transport has rarely recurred on such a scale and so its attraction for private firms has diminished. The supply of transport has gradually been assumed by the state or by a small number of large firms, usually heavily subsidized by the central government. Under these conditions different decision-making criteria apply to the production of transport and a different spatial structure emerges.

The relationship between demands for movement and the evolving spatial form of the transport response is a popular topic in geographical studies. Given the dynamic two-way relationship between transport and spatial readjustment it is hardly surprising that no definitive model of transport development has emerged as yet. Janelle (1969) has attempted to conceptualize the iterative process of transport expansion and spatial readjustment, and some (e.g. Kansky 1963; Kolars and Malin 1970) have gone so far as to attempt to simulate the spatial development of transport networks. An oft-quoted geographical study of transport network development is an attempt to characterize the network expansion of less-developed countries in an idealized four-stage model (Taaffe, Morrill and Gould 1963):

(1) Development of scattered coastal ports
(2) Selective growth of internal lines of penetration
(3) Growth of interconnections between nodes on lines of penetration
(4) Evolution of high priority links based on agglomeration economies at selected internal nodes and ports

This staged approach has been criticized on the grounds that network expansion is continuous, but its authors point out that the second stage is fundamental to the transport geography of a less-developed country and that the lumpiness of transport investment and the jerky process of unbalanced development (see above, pp. 148–150) add viability to the stages. More descriptive studies of the close relationships between transport form and function include Hilling's (1969) account, based once more on a number of stages, of the changing number, location, size and infra-structure of West African ports in response to the changing nature of their throughput; and Smith's (1964a) account of the stages in the growth of the transport network of New South Wales. A sequence of port development has been idealized by Bird (1963) and applied to ports in Britain, Australia (Bird 1963; 1968) and East Africa (Hoyle 1968).

The spatial structure of networks

Transport networks are spatial structures designed to channel flows from points of demand to points of supply and so to link these points together in a transportation system. The previous section has shown that it is difficult to isolate and define transport demand and supply but, by divorcing the function of transport networks from their inert spatial form, it is possible to derive some very useful descriptive indices (Garrison 1960; Kansky 1963).

Any transport network may be considered as a topologic graph with three parameters from which quantitative measurements may be computed as a basis for the objective description, comparison and evaluation of networks. These parameters are (i) the number of separate (non-connecting) sub-graphs in the network (G); (ii) the number of links (or edges) in the network (E); and (iii) the number of nodes (or vertices) in the network (V). Nodes may be the origins or destinations of flows and they are points at which flows in the network can change their volume, direction of movement and mode of transport. They range in size and complexity from a road junction to a major international port but they are all equipped with the physical infra-structure to enable the various types of flow adjustment to take place and so may become attractive locations for firms which can cut out one set of handling charges by locating adjacent to the transport intersection.

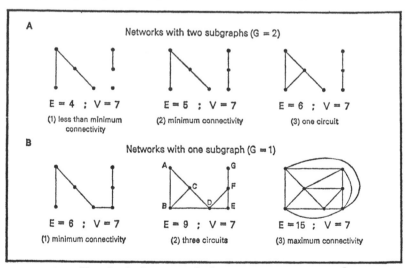

9.2 *Topological maps of planar transport networks.*

Fig. 9.2 shows a series of networks in which, for the sake of arithmetic simplicity, the number of subgraphs is limited to two (G = 2) and the number of nodes is held constant (V = 7). Furthermore, all the networks shown in fig. 9.2 are planar graphs in that they are confined to a two-dimensional surface so that none of the links can intersect without creating a node. This characteristic is typical of most land-based modes of transport, although some motorway flyovers may contradict this generalization. It is atypical of the three-dimensional criss-crossing

TABLE 9.1 *Some quantitative measures of planar transport networks*

$$\text{Beta index} \quad = \quad \frac{\sum\limits_{G=1}^{n} E}{\sum\limits_{G=1}^{n} V}$$

$$\text{Cyclomatic number} = \quad \sum\limits_{G=1}^{n} E - \sum\limits_{G=1}^{n} V + G$$

$$\text{Alpha index} \quad = \quad \left(\frac{\sum\limits_{G=1}^{n} E - \sum\limits_{G=1}^{n} V + G}{2 \sum\limits_{G=1}^{n} V - 5G} \right) 100$$

$$\text{Gamma index} \quad = \quad \left(\frac{\sum\limits_{G=1}^{n} E}{3(\sum\limits_{G=1}^{n} V - 2G)} \right) 100$$

of aircraft flight paths and of the intersection of modally distinct surface networks, except at specially constructed nodes (e.g. ports, railway stations) designed to facilitate modal-integration. Descriptive indices are also easily calculated for such non-planar networks, although table 9.1 lists only some of the planar indices that may be constructed from the parameters of the planar networks shown in fig. 9.2. The results of these calculations are presented in table 9.2.

TABLE 9.2 *Topological indices for networks mapped in fig. 9.2*

	A			B		
	(i)	(ii)	(iii)	(i)	(ii)	(iii)
Beta index	0·57	0·71	0·86	0·86	1·29	2·14
Cyclomatic number	−1	0	1	0	3	9
Alpha index	<0	0	25%	0	33%	100%
Gamma index	44%	55%	66%	40%	60%	100%

The indices defined in table 9.1 are primarily concerned with measuring the degree of connection, or connectivity, between all the nodes in the network. The connectivity of a network is a particularly important measure because it embodies an element of evaluation which may be applied to cross-sectional studies of networks in different places or time-series analyses of the development of a network in one area.

The beta index is a simple measure of connectivity in terms of the average number of links per node within the network:

$$\frac{\sum\limits_{G=1}^{n} E}{\sum\limits_{G=1}^{n} V}$$

Values of this index range from zero to three (Haggett 1965, 238). A value of zero shows that no network exists and higher values result from increasingly complex networks. A minimally connected network is defined as one containing no isolated nodes and with

$$\sum_{G=1}^{n} V - G \quad \text{links.}$$

Thus a modified version of the beta index:

$$\frac{\sum\limits_{G=1}^{n} E}{\sum\limits_{G=1}^{n} V - G}$$

would give a value of 1 for a minimally connected network; higher values of the beta index result from increasingly complex and connected networks.

The addition of one link to a minimally connected network would increase its connectivity and, unless the link connected two minimally connected subgraphs, would introduce network circuitry as a circuit would be created. A circuit is a closed path in which the initial node is the same as the terminal node. The number of circuits within a network is given by the cyclomatic number. Wherever circuits exist the number of links must exceed

$$\sum_{G=1}^{n} V - G$$

because this number defines a minimally connected network.
The number of circuits may be calculated by subtracting this minimal number of links from the actual number of links:

$$\sum_{G=1}^{n} E - \left(\sum_{G=1}^{n} V - G \right)$$

which, multiplying out, may be rewritten as:

$$\sum_{G=1}^{n} E - \sum_{G=1}^{n} V + G$$

which is the cyclomatic number.

In any planar network with more than two nodes the addition of one extra node increases the maximum number of links by three. Thus the maximum number of edges may be expressed as:

$$3\left(\sum_{G=1}^{n} V - 2G\right)$$

The maximum number of circuits within any network may be found by subtracting the number of links in a minimally connected network from this number:

$$3\left(\sum_{G=1}^{n} V - 2G\right) - \left(\sum_{G=1}^{n} V - G\right) = 2\sum_{G=1}^{n} V - 5G$$

The alpha index is a measure of the ratio between the observed number of circuits (the cyclomatic number) and the maximum number of circuits that may exist in the network:

$$\left(\frac{\sum_{G=1}^{n} E - \sum_{G=1}^{n} V + G}{2\sum_{G=1}^{n} V - 5G}\right)$$

The range in values of the alpha index, which measures the degree of network circuitry, runs from zero with no circuits, to one when the actual number of circuits is equal to the maximum number, and so the index may be expressed as a percentage.

The gamma index is rather more simple, measuring the ratio of the observed number of links and the maximum number of links in any network:

$$\left(\frac{\sum_{G=1}^{n} E}{3\left(\sum_{G=1}^{n} V - 2G\right)}\right)$$

Again the range in value of the gamma index ranges from zero, with no links, to one in which, with the exception of links that would intersect, every node in the network has a link connecting it to every other node. Thus it may be expressed as a percentage.

All the indices described so far are concerned with the network as a whole, although network analysis can yield valuable measures of the accessibility of individual nodes. One such measure is derived from the connectivity matrix (fig. 9.3) which represents the links between the nodes of a network in matrix form. A figure of one in the matrix denotes the presence of an inter-nodal link, a zero denotes the absence of such a link. The distance between pairs of nodes is expressed as the number of intervening links along the shortest path connecting them. The row sum for each node is a measure of its accessibility in terms of this measure of distance and the grand total provides a measure of the total size of the network in terms of its total number of links. This measure is known as the dispersion value of the graph. The average path length in the network is obtained by dividing the row sum by the total number of positive values in the row. Fig. 9.3 reveals that C is the most accessible node and that E is the most inaccessible.

The location of networks

The total costs of a network over a given period of time are made up of fixed construction costs and variable movement and maintenance costs. This combination of fixed and variable costs results in a 'U' shaped average total cost curve. As long as the variable marginal costs generated by increasing flows over the network are less than average total costs, the average total cost curve, which includes both variable and fixed costs, will fall, but when the marginal costs begin to rise as a result, say, of congestion or increased maintenance charges, then the average total cost curve will rise (cf. fig. 5.4, p. 71).

Fixed construction costs are closely determined by the spatial design of networks as they are a function of total network length. But variable flow volumes are a function of the size of the pre-existing effective demand for movement and the extent to which the spatial design of networks is able both to tap this demand and to concentrate the flows over a limited number of links in order to achieve economies of scale. The balance between fixed and variable costs is, therefore, critical in determining the spatial pattern of networks (Haggett and Chorley 1969) and intrudes yet another variable into the transport decision-makers' already complex investment problem.

Two types of node characterize most networks: fixed nodes which are the origins and destination of flows and floating nodes which may be created by the operation of the transport system in servicing flows between the fixed nodes. In fig. 9.4 three fixed nodes are joined by a straight line (a), a pair of direct lines with a vertex at E (b), a set of three straight lines radiating from a vertex, which forms a route junction or node (c), and by three direct links forming a triangular network (d). In this simple case the locational structure of the network depends partly upon the spatial arrangement of the fixed nodes: (a) differs from (b) in this respect alone, and upon the oppor-

9.3 *Connectivity matrix.* (*Source: Haggett 1972.*)

.to	A	B	C	D	E	row sums	average path length
from							
A	0	1	1	2	3	7	1.75
B	1	0	1	2	3	7	1.75
C	1	1	0	1	2	5	1.25
D	2	2	1	0	1	6	1.50
E	3	3	2	1	0	9	2.25
					total	34	1.70

tunity cost of providing the network. In (b) a more direct connection between D and F would have increased the length of the network, as a siding would have been necessitated to serve E, and so would have

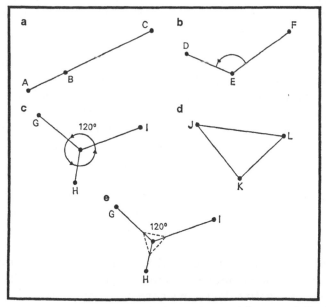

9.4 *Costs and network location.* (*Source: Abler, Adams and Gould 1971.*)

increased total costs. By contrast in (c) direct connection between the three nodes would have produced a longer and more expensive network than that radiating from a floating node. As a simple generalization if any of the interior angles of the triangle created by the three points is greater than 120° the cheapest (shortest) link between those points is represented by direct connection as in (a). If any angles are equal to or less than 120° the cheapest solution is to create a floating node with routes radiating out to the other nodes at appropriate intervals (c and e).

The minimization of construction costs may serve to increase user costs. In fig. 9.4(d) user costs are minimized as each point is linked directly to another, whereas in (e) builder costs are minimized, although the development of short-cuts (shown by the dotted lines) represents one approach to a compromise. Several floating nodes may be required when applying the 120° degree rule to more than three fixed nodes and the problem of minimizing the total length of the network is increased because each floating node must be located simultaneously to create the optimum network. Some solutions to this problem are reviewed by Haggett and Chorley (1969, 211–17).

Variables other than the spatial arrangement of the nodes and the costs deriving from the structural design of networks enter into the decisions on network locations. Although the shortest line between two points on a plane is a straight line (a route which would minimize both builder and user costs) positive and negative route deviations can be induced (Haggett 1965). Positive deviations (lengthening the route to serve more places) are illustrated in fig. 9.5A, which shows four alternative solutions (from the 64 possible) to the problem of linking two terminal and six intervening cities. The potential revenue deriving from each city is shown and the cost of the transport link to the city is assumed to be directly proportional to its length. If the aim is to maximize net return, solution (c) might be chosen, but social considerations might result in (d), which would give access to two more cities but would also reduce the net return. Many other solutions involve the use of floating nodes and branch lines but the cost of providing junction facilities at the floating nodes would need to be balanced against the savings resulting from reduced route mileage. Furthermore, as transport costs exhibit scale economies, the direct extra revenue which results from the extension of the railway may be supplemented by a lowering of unit costs resulting from the increased use of the network as a whole.

Negative deviations result from the enforced extension of routes to avoid costly environmental barriers to straight-line routing. The spatial implications of negative deviations have been demonstrated by Werner (1968) for the construction of a route between two points separated by

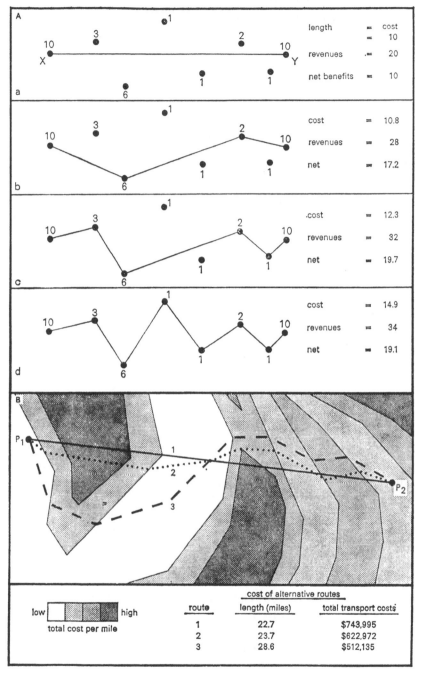

9.5 *Positive and negative network deviation. (Sources: A. Abler, Adams and Gould 1971; B. Werner 1968.)*

a variety of environmental cost conditions (fig. 9.5B). The least-cost route shows a marked deviation from the straight-line cost and demonstrates a preference for extended location in the least-cost areas. Negative deviations may also be induced by factors such as the desire to retain national sovereignty over routes which consequently deviate from a straight line in order to stay on the right side of a political boundary. Similarly political boundaries may act as barriers to transport investment or hinder supra-national cooperation in transport investment.

The shape of transport networks is clearly an important extension of the simplified topological characteristics discussed in the previous section. A measure of the shape effect of positive and negative deviations – the route factor – has been proposed by Kansky (1963) and may be expressed as:

$$S = \frac{[\sum_{i=1}^{n} (O-E)^2]}{V}$$

where S = route factor
 O = observed distance between two nodes
 E = straight line ('desire line') distance between two nodes
 V = number of nodes in the network.

The consideration of circuitous networks as the result of positive and negative deviations adds a degree of generality to the study of devious transport links (e.g. Appleton, 1965a; Warntz 1961), and underlines an important point: that the geographic locations occupied by routeways are created by economic evaluation and a process of decision making (Meinig 1962) rather than by a process of natural endowment.

The quantitative measurement of network location has received more attention from geographers at the regional level than at the level of the individual route. Network density per unit area is normally used as the dependent variable and studies at several scales reveal that within individual cities (e.g. Borchert 1961; Owens 1968), nation states (Taaffe, Morrill and Gould 1963; Kolars and Malin 1970) and at the world level (e.g. Berry 1960; Garrison and Marble 1961; Kansky 1963) there is a positive association between it and variables like population density and size of G.N.P. Such variables act as surrogates for the volume of circulation or demand for movement. On the other hand, there is normally a negative relationship between environmentally costly zones and the provision of transport networks (e.g. fig. 9.5B).

An attempt to review and fuse the statistical relationships between road and rail transport systems and their socio-economic and physical environment (Haggett 1967) shows that the various structural and den-

sity variables within the rail sub-system are much more closely inter-
locked, especially around the density measure, than in the road
sub-system. Furthermore, although the links between environmental
and transport systems are weaker than those within the transport
system itself, the railway density measure is most closely associated
with Berry's (1960) descriptive but complex technological factor of
economic development. The rail systems appear to be far more closely
influenced by the economy and the physical environment of the
countries in which they are set than are the road systems.

Conclusions

In this chapter we have considered transport as a spatial system, the
outcome of a complex and iterative process of reconciliation between
demand and supply. The complexity of this relationship derives, in
large measure, from the externalities associated with transport and
herein lie some of the most fascinating questions associated with the
spatial analysis of transport. For example, two important fields of
inquiry have received scant attention: the imprecisely understood role of
transport in the redistribution of economic development between areas
and of income between individuals and, despite the general decline of
transport costs for individual firms, the complex relationship between
the spatial structure of an economy and the amount of resources
devoted to transport provision. The recent and welcome appearance
of two texts in economic geography dealing specifically with transport
(Taaffe and Gauthier 1973; Hay 1973) will hopefully stimulate further
work in these fields.

Part 4

The economy: growth and development

Economic growth is difficult to initiate and to sustain. It is increasingly under attack as a fundamental aim of economic policy because of the ecological threat that it poses and its ability via technical innovation to dehumanize the work situation. However, the spatial distribution of the distaste for growth appears to be positively correlated with levels of economic development. The observed tendency within capitalist economies to concentrate economic growth in selected, narrowly defined locations and the resistance of the rich to attempts to redistribute their economic power to the poor, appear to provide two good reasons for the adoption of growth policies which may facilitate the task of income redistribution. But the problems of socio-economic inequalities are complex. Direct attempts to deal with them may founder unless the underlying sources of economic power, including the economic, social, political and cultural values of society which sanction inequalities of economic power, are also controlled. Some of these issues are discussed in this section which also attempts to pull the threads of the preceding discussion together by focussing attention upon two major and related problems of the world economy.

10 Economic growth and development

It is only recently that geographers have become seriously interested in the complex phenomena associated with the development of economies. There are at least three interconnected reasons for this. To some extent, as Keeble (1967) has noted, the lack of interest among geographers in the problems of economic change was due to the earlier idiographic approach in which emphasis on the uniqueness of areas made theoretical generalizations difficult to reach, whereas the present nomothetic approach encourages the geographer to search for universal laws or principles. Secondly, it is clear that there has until recently been a tendency among economic geographers to adopt approaches that are descriptive rather than explicitly theoretical, whereas the study of development requires a rigorous theoretical base. Thirdly, there has for long been an easily accepted assumption that development within economies, at whatever scale, is largely or even exclusively the concern of economists. But it is now generally recognized – by economists as well as non-economists – that study of this kind of change demands an approach that is essentially interdisciplinary. Development is not simply an economic process and the practical problems it raises never fall neatly within the confines or competence of any one discipline.

This is apparent as soon as any attempt is made to define development. For some writers, development and growth are synonymous terms, it being argued that to make a distinction between them implies a number of value judgments about the nature of economic and social change. Others find it useful to make a distinction between growth (associated with advanced economies) and development (associated with economically backward economies), although such a distinction seems to perpetuate the misleading notion of a clear disjunction between the developed and less-developed world. In the present study, however, the term 'growth' refers solely to the aggregate and strictly economic or material improvement in an economy. 'Development' is taken to refer to a much wider range of variables, including especially that whole social, economic and political process which results in a perceptible and cumulative rise in the standard and 'quality' of life for an increasing proportion of the population.

Aims and choices

It has been necessary on a number of occasions to refer to the aims of directed change within economies. In general terms a common goal is to raise per capita incomes, but personal and spatial inequalities in per capita incomes between regions of a country, between countries and, more broadly, between the less-developed and developed countries raise one of the major problems of the world economy. It is true that many writers draw attention to the limitations of the available standards of comparison between regions, states and groups of states and Berry's (1960) work drew early and specific attention to the need to think more in terms of a continuum in levels of development – however defined – than in terms of 'developed and developing' or 'tropical and non-tropical' countries, as if they were mutually exclusive categories. It is also true that some of the major variables in assessing levels of development are not susceptible to rigorous quantitative analysis. But, more important, the general statement that all countries are aiming to raise per capita incomes grossly oversimplifies the issue for those countries where the decisive criteria of 'successful development' are more social, political or strategic than purely economic or material. Thus whereas in one country the overriding objective is a high rate of growth of income, in another it may be full employment, the development of backward regions, the creation of strategic industries, the reduction of reliance on foreign trade, or the furtherance of a socialist society.

One of the implications of considering these various possible aims of development and, moreover, of following the approach adopted in these pages, is that sooner or later one comes up against the need to consider the planning choices open to decision makers as to the nature and rate of development. Most commonly, of course, this is seen to operate at the national level, although of equal importance may be the development of an economy at local, urban, regional or international levels. In all cases, decisions have to be made about the allocation of resources over time and space, about the techniques used, whether to follow or reject balanced growth, and about the relative importance of agriculture, industry, education, health, transport or irrigation in the planning process. Furthermore, any decision-making body has to establish its priorities and strategy. Most frequently it is quickly recognized that a conflict can arise between the interests of total or aggregate economic growth, and the interests of economic equality between classes, groups or regions. Economic and environmental interests may also conflict, as may the interests of the individual and those of the larger groups of which he is a member; and this latter difficulty raises the whole question of scale and value systems in development studies (Gilbert 1971; Connell 1971; Brookfield 1973).

Planning for economic change represents an attempt to look at the various lines of development within a specific area, not as individual strands or projects but as part of the whole problem. Integration and balance of the various elements in an economy are normally important ingredients of any development plan. Plans, at whatever scale, represent an attempt at integrated and dynamic analysis within the framework of a particular areal setting. And this is why the examination of development plans and problems often provides the economic geographer with a useful focussing or integrating point in his analysis.

Planning for development

The role of government in decision making is particularly well demonstrated in development planning analyses. Any development plan, at whatever scale – whether international, national or regional – involves the taking of the kinds of decisions discussed in preceding chapters.

A development plan today, assuming it to have proceeded beyond the stage of a simple list of projects, normally looks at the economy as a whole, the private as well as the public sectors, and is usually a comprehensive plan in that the public expenditure programme is fitted into the framework of the expected overall development of the economy. This implies an element of macro-economic forecasting, making projections of the expected movements of the economy. This forecasting is, of course, based on a detailed evaluation of projects within the public sector, although such an analysis may not always be feasible, more particularly in less-developed countries. Nevertheless, a development plan commonly aims deliberately at influencing the private sector, either by direct aid to assist private investment or by setting out within the plan what the government hopes to be able to provide in terms of infra-structure, skilled manpower, and new projects, and what the private sector taken as a whole could do. This may give the private sector the confidence to carry out the higher levels of investment required (Ord and Livingstone 1969). In a broader sense, too, development planning can assist in the task of nation-building. This is probably true in China and Tanzania where the wider picture of the egalitarian type of society being aimed at is presented.

The benefits of development planning stem largely from the better evaluation of projects against a background of anticipated development of the economy as a whole; from the long-term view required by some decisions, for instance by providing training facilities for skilled manpower in the future; and from the complementary nature of investment decisions – the viability of one project may be conditional upon another project being undertaken. Benefits also accrue from the necessity for overall balance in an economy and from the possibility of using the plan to stimulate effort. Finally, development planning may be vital to the presentation of a case for foreign aid!

On the other hand, development planning has very clear limitations. One of the most basic is the lack of sufficiently accurate or detailed information on the economy as a whole and about specific projects. 'The usefulness of the sophisticated mathematical exercise underlying success in development plans is entirely dependent upon the quality of the statistics fed into the calculations' (Ord and Livingstone 1969, 432). Other limitations include the natural tendency of development plans to over-reach themselves, together with the fact that failure to achieve targets can produce loss of faith by the public and so jeopardize future planning. Again, development planning cannot by itself produce projects. Another limitation is that where there is a large private sector, this is not directly controlled by the planning authority. Similarly, all economies lie outside the control of governments in certain respects – where there are export-dependent economies or economies which rely heavily on foreign aid, foreign enterprise or foreign capital and where the local market is insufficient to support new industries. But perhaps the most crucial limitation of all may lie in the absence of any adequate machinery or administrative organization – at the local, regional or national levels – with which to implement planning.

Models of economic development
In the study of planning for development, whether in developed or less-developed countries, economic geographers have shown a particular interest in the application of generalized models. However, relatively little attention has been paid to the application of non-spatial models, and this tendency is explicable in terms of the widely held view that the economic geographer's interest in the processes and planning of development should be restricted to spatial analyses. It is frequently argued that the study of economic development by economists is over-theoretical, obsessed with sectoral rather than spatial analyses, and out of touch with the realities of specific development situations. Consequently, many economic geographers concern themselves exclusively with spatial aspects of planning or development, thereby committing the same sin of arbitrary selectivity and incompleteness that they complain of in economists. For other economic geographers, however, the need to study process as well as form and to include in their analyses everything that seems to be relevant to the applied problems under examination, means that non-spatial models may well be as significant as spatial models. Certainly some acquaintance with the major non-spatial models of development seems necessary. These include stage models, notably that of Rostow (1960) on which a good deal of literature now exists. The thesis of Clark (1960) and Fisher (1952) is also of

particular interest in that it focusses attention on sectoral shifts within the structure of an economy as it develops. The attraction of all such historically based or stage theories is that they provide coherent frameworks for an understanding of the regional, national and continental spatial variations in prosperity and levels of economic development.

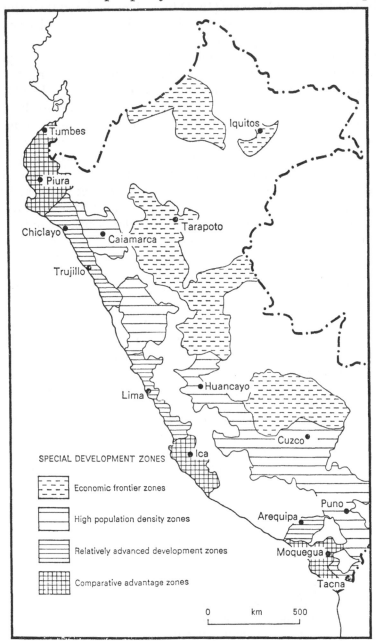

10.1 *Regional economic development in Peru. (Source: Slater, forth-coming.)*

It is probably true to say, however, that this historical-stage approach to the study of development has attracted more criticism than support in the recent literature (Streeten 1971).

The interest shown by economic geographers in the spatial patterns and models of development within and between economies is increasing rapidly. Chisholm and Manners (1971) have demonstrated the value and relevance of this kind of approach in their study of the British economy. For a developing country Bradnock (forthcoming) has discussed the case of India, where regional inequalities in levels of development appear to have grown rather than declined, in spite of planning efforts to achieve the opposite. At the theoretical level, both Myrdal (1957) and Hirschman (1958) have argued that in the early stages of development, and before government has intervened, regional inequalities in income will tend to increase. Furthermore, in Myrdal's model the flow of new commodities from factories to peripheral regions introduces competition which traditional industries cannot withstand – the 'backwash effect'. In India, regional concentration of industrial productive capacity has certainly increased. The major metropolitan centres have been the foci of much of the increase in manufacturing employment. The only effective counter-magnet to these centres has been raw material deposits, notably of iron ore and coal, which have pulled the location of heavy industries away from the largest cities. Direct employment in the basic industries, however, is small. This contrasts with experience in the United States where Isard and Kuenne (1953) found that the location of a new iron and steel plant in the Greater New York–Philadelphia region had a dramatic multiplier effect on employment. In France Perroux (1961) and, in Peru, Slater (forthcoming) have discussed the regional zoning policies of economic planning. Slater has shown how concentration into poles and axes of development is occurring in spite of attempts to reverse the trend (fig. 10.1).

Regional studies of income elucidate this problem. Williamson (1965), in an empirical survey of international patterns of regional inequality, has calculated indices of interregional inequality in 24 countries, 20 of them in the middle or upper income range. The index V_w is a weighted index of regional inequality, and is calculated from the formula:

$$V_w = \sqrt{\dfrac{\sum\limits_i (y_i - \bar{y})^2 \dfrac{f_i}{n}}{\bar{y}}}$$

where f_i = population of the ith region
n = national population
y_i = income per capita of the ith region
$\bar{y}$ = national income per capita

The resulting values range from a low of 0·058, indicating a low degree of inequality, to 0·700, indicating a great difference in income.

Williamson's study suggests that regional incomes first diverge and then converge as *per capita* incomes rise. However more recent data for other underdeveloped countries fail to support this conclusion. Given the structural difficulties of national economic development it is possible that *per capita* incomes may not rise to the levels thought to be associated with the process of regional convergence. Furthermore regional inequalities tend to intensify with growth despite state-sponsored regional development programmes which may themselves deter capital and so further retard national growth. In short Williamson's proposition of income divergence followed by convergence 'should be regarded with serious reservation' (Gilbert and Goodman, 1976, 135).

In all these models of development, spatial or non-spatial, regional or national, it is worth emphasizing that they help simply to guide and structure logical thought rather than provide mirrors of reality or prescriptions for planning. Certainly as far as development planning is concerned, our lack of understanding of the most relevant processes, our inability to handle efficiently even the available data, and the subjective nature of most of the decision making, give support to the contention that 'there is no body of scientific principles to be used in making development plans' (Lewis 1959, 417) and that in the final analysis commonsense and experience are the only tests one can apply to its assessment.

Limits to growth

One of the most recent trends in the study of development has been the re-examination, in the context of the ecological problems mentioned earlier on pp. 9–10, of the assumptions commonly held about development. There is increasing concern for what is termed the 'quality of life' rather than the level of material provision. In many countries – and especially in those countries, like Japan, which are experiencing rapid growth rates – the assumption that economic growth is necessarily a desirable end is now seriously questioned by many writers: higher per capita incomes, more consumer goods and greater material prosperity in all its manifestations do not, so the argument goes, necessarily add up to greater human satisfaction and happiness. Certainly the ecological implications of unrestricted growth are now generally understood. The publication of the *Blueprint for Survival* (Goldsmith *et al.* 1972) in Britain, the work of the Club of Rome, published in *Limits to Growth* (Meadows *et al.* 1972) and the work of the Stockholm Conference (1972) have focussed attention on the dangers of unrestricted growth in the world economy. The Club of Rome's analysis uses a formal, written model of the world, built specifically to investigate five major trends of global concern. With this model, the authors attempt to understand the causes of these trends, their interrelationships, and their implications as much as one hundred years in the future (fig. 10.2). The conclusions reached are three. First, if the present trends of growth in world population, industrialization, pollution, food production and resource depletion continue unchanged, the limits to growth will be

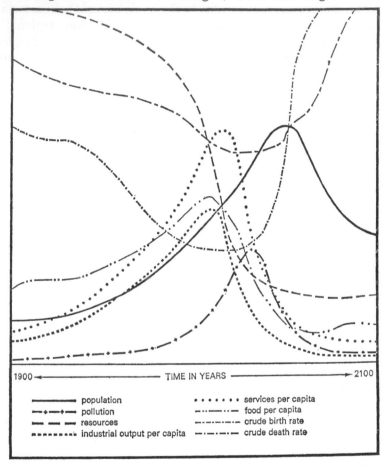

1900 ◄─────────────── TIME IN YEARS ───────────► 2100

──────── population   • • • • • • services per capita
─•──•──•─ pollution   ─•••─•••─•••─ food per capita
─ ─ ─ ─ resources   ──•─••─•─ crude birth rate
••••••••• industrial output per capita   ─•─••─••─ crude death rate

10.2 *Economic growth and some socio-ecological trends. (Source: Meadows et al. 1972.)*

reached sometime within the next one hundred years. The most probable result will be a rather sudden and uncontrollable decline in both population and industrial capacity. Secondly, it is still possible to alter these growth trends and to establish a condition of ecological and economic stability that is sustainable far into the future. The state of global equilibrium could be designed so that the basic material needs of each person on earth are satisfied and each person has an equal opportunity to realize his individual potential. Thirdly, if the world's peoples decide to strive for the second outcome rather than the first, the sooner they begin working to attain it, the greater will be their chance of success (see Simmons 1973).

These arguments and conclusions have, predictably, generated a good deal of criticism and scepticism. In the less-developed countries, or in the poorer regions of the developed world, the implications of the *Limits to Growth* arguments are unlikely to gain much support. It is understandably doubted in poor countries whether the policy of global equilibrium is ever likely to include in practice the rapid increase of wealth in the poorer parts of the world. This in spite of the Club of Rome's statement that 'world equilibrium can become a reality only if the lot of the so-called developing countries is substantially improved, both in absolute terms and relative to the economically developed nations' (*ibid.*, 191). In the developed countries, criticisms of the kind of thinking implied in the Limits to Growth philosophy draw attention to the limited number of variables analysed, to the impossibility of predicting accurately the trend in any one variable, yet alone the complex of variables, and to the unrealistic nature of some of the assumptions made in the analyses.

Conclusions

There seems little doubt that the study of development will increasingly be accepted as a focus for the examination of the economy by geographers. From a methodological point of view such study is clearly of particular relevance to the approach followed in these pages in that it implies a look at the processes of change within economies. Furthermore, the analysis of development in this context relates not so much to change within single elements in an economy as to the total, integrated complex of elements which characterize all economies, at all scales and at all levels. This complexity makes inevitable and desirable the increasing specialization of intellectual activity within the field of economic geography. But in an applied or practical sense, developmental considerations are crucial to any understanding or solution of the two problems which have been identified time and again in the pages of this book. First, there are the excessive structural and spatial inequalities and related social injustices, both within and between economies. These inequalities are becoming more rather than less apparent at almost every scale. Capitalist economies are controlled, quite as much as are command economies, by a small number of their most powerful elements. Furthermore, this inequality of control is encouraged in capitalist economies, the institutions of which are successfully postponing any real attempt to redistribute economic power more equitably (Urry and Wakeford 1973). A recognition of the institutional

G

acceptance and allocation of power and the ways in which it is used must form the basis of any conscious attempt to change the real world economy. Secondly, attention has been directed to the ecological problems associated with rapid economic growth. And here, too, the socio-cultural and political contexts have been shown to be crucial, even decisive elements in their control. Taken together, these related problems are perhaps the most urgent and intractable facing mankind today and, as such, must be central to the analysis by economic geographers of the real world.

Appendix

The metric system; conversion factors and symbols

In common with several other text book series *The Field of Geography* uses the metric units of measurement recommended for scientific journals by the Royal Society Conference of Editors.* For geography texts the most commonly used of these units are:

Physical quantity	Name of unit	Symbol for unit	Definition of non-basic units
length	metre	m	basic
area	square metre	m^2	basic
	hectare	ha	$10^4 m^2$
mass	kilogramme	kg	basic
	tonne	t	$10^3 kg$
volume	cubic metre	m^3	basic-derived
	litre	l	$10^{-3} m^3$, $1\ dm^3$
time	second	s	basic
force	newton	N	$kg\ m\ s^{-2}$
pressure	bar	bar	$10^5\ Nm^{-2}$
energy	joule	J	$kgm^2 s^{-2}$
power	watt	W	$kgm^2 s^{-3} = Js^{-1}$
thermodynamic temperature	degree Kelvin	°K	basic
customary temperature	degree Celsius	°C	$t\ °C = T\ °K - 273\cdot15$

Fractions and multiples

Fraction	Prefix	Symbol	Multiple	Prefix	Symbol
10^{-1}	deci	d	10	deka	da
10^{-2}	centi	c	10^2	hecto	h
10^{-3}	milli	m	10^3	kilo	k
10^{-6}	micro	μ	10^6	mega	M

* Royal Society Conference of Editors, *Metrication in Scientific Journals*, London, 1968.

The gramme (g) is used in association with numerical prefixes to avoid such absurdities as mkg for μg or kkg for Mg.

Conversion of common British units to metric units

Length

1 mile	= 1·609 km	1 fathom	= 1·829 m
1 furlong	= 0·201 km	1 yard	= 0·914 m
1 chain	= 20·117 m	1 foot	= 0·305 m
		1 inch	= 25·4 mm

Area

1 sq mile	= 2·590 km²	1 sq foot	= 0·093 m²
1 acre	= 0·405 ha	1 sq inch	= 645·16 mm²

Mass

1 ton	= 1·016 t	1 lb	= 0·454 kg
1 cwt	= 50·802 kg	1 oz	= 28·350 g
1 stone	= 6·350 kg		

Mass per unit length and per unit area

1 ton/mile	= 0·631 t/km	1 ton/sq mile	= 392·298 kg/km²
1 lb/ft	= 1·488 kg/m	1 cwt/acre	= 125·535 kg/ha

Volume and capacity

1 cubic foot	= 0·028 m³	1 gallon	= 4·546 l
1 cubic inch	= 1638·71 mm³	1 U.S. gallon	= 3·785 l
1 U.S. barrel	= 0·159 m³	1 quart	= 1·137 l
1 bushel	= 0·036 m³	1 pint	= 0·568 l
		1 gill	= 0·142 l

Velocity

1 m.p.h.	= 1·609 km/h
1 ft/s	= 0·305 m/s
1 U.K. knot	= 1·853 km/h

Mass carried × distance

1 ton mile = 1·635 t km

Force

1 ton-force	= 9·964 kN
1 lb-force	= 4·448 N
1 poundal	= 0·138 N
1 dyn	= 10⁻⁵ N

Pressure

1 ton-force/ft²	= 107·252 kN/m²
1 lb-force/in²	= 68·948 mbar
1 pdl/ft²	= 1·488 N/m²

Energy and power

1 therm	= 105·506 MJ
1 hp	= 745·700 W(J/s) = 0·746 kW
1 hp/hour	= 2·685 MJ
1 kWh	= 3·6 MJ
1 Btu	= 1·055 kJ
1 ft lb-force	= 1·356 J
1 ft pdl	= 0·042 J
1 cal	= 4·187 J
1 erg	= 10^{-7} J

Metric units have been used in the text wherever possible. British or other standard equivalents have been added in brackets in a few cases where metric units are still only used infrequently by English-speaking readers.

References

ABLER, R., J. S. ADAMS and P. GOULD (1971) *Spatial organization: the geographer's view of the world*. Englewood Cliffs, N.J.

ADAMS, J. G. U. (1970) *The spatial economy of West Africa*. Unpubl. Ph.D. thesis, Univ. of London.

ALEXANDER, J. W. (1963) *Economic geography*. Englewood Cliffs, N.J.

ALEXANDERSSON, G. (1967) *Geography of manufacturing*. Englewood Cliffs, N.J.

ALONSO, W. (1964) *Location and land use: toward a general theory of land rent*. Cambridge, Mass.

AMBROSE, P. J. (1968) An analysis of intra-urban shopping patterns. *Tn Plan. Rev.* 39, 327–34.

ANDREWS, H. F. (1971) Consumer behaviour and the tertiary activity system, in A. G. WILSON (ed.) *Urban and regional planning*. London.

ANDREWS, H. F. (1973) Urban structure correlates of tertiary activity. *Reg. Stud.* 7, 263–70.

APPLEBAUM, W. (1954) Marketing geography, in C. JONES and P. JAMES (eds.) *American geography – inventory and prospect*. Syracuse.

APPLETON, J. H. (1962) *The geography of communications in Great Britain*. London.

APPLETON, J. H. (1965a) *A morphological approach to the geography of transport*. Hull.

APPLETON, J. H. (1965b) Historical geography and the Beeching Report. *Scot. Geogr. Mag.* 81, 38–47.

BACON, R. W. (1971) An approach to the theory of consumer shopping behaviour. *Urb. Stud.* 8, 55–64.

BAKER, O. E. (1921) The increasing importance of the physical conditions in determining the utilization of land for agricultural and forest production in the United States. *Ann. Assoc. Am. Geogr.* 11, 17–46.

BAKER, O. E. (1925) The potential supply of wheat. *Econ. Geogr.* 1, 15–52.

BARAN, P. A. and P. M. SWEEZY (1968) *Monopoly capital*. Harmondsworth.

BASSETT, K. and P. HAGGETT (1971) Towards short-term forecasting for cyclic behaviour in a regional system of cities, in M. CHISHOLM, A. E. FREY and P. HAGGETT (eds.) *Regional forecasting*. Proc. of the twenty-second symposium of the Colston Res. Soc. London.

BAYLISS, B. T. and S. L. EDWARDS (1970) *Industrial demand for transport*. H.M.S.O., London.

BECKERMAN, W. (1956) Distance and the pattern of intra European trade. *Rev. Econ. and Statistics*, 38, 31–40.

BECKMAN, T. N. and N. H. ENGLE (1937) *Wholesaling, principles and practice*. New York.

BERRY, B. J. L. (1959) Recent studies concerning the role of transportation in the space economy. *Ann. Assoc. Am. Geogr.* 49, 328–342.

BERRY, B. J. L. (1960) An inductive approach to the regionalization of economic development, in N. GINSBURG (ed.) *Essays on geography and regional development*. Univ. of Chicago, Dept. Geogr. Res. Pap. 62, Chicago.

BERRY, B. J. L. et al. (1966) *Essays on commodity flows and the spatial structure of the Indian economy*. Univ. of Chicago, Dept. Geogr. Res. Pap. 111, Chicago.

BERRY, B. J. L. (1967) *Geography of market centres and retail distribution*. Englewood Cliffs, N.J.

BERRY, B. J. L., H. G. BARNUM and R. J. TENNANT (1962) Retail location and consumer behaviour. *Pap. and Proc. Reg. Sci. Assoc.* 9, 65–106.

BERRY, B. J. L. and F. E. HORTON (1970) *Geographic perspectives on urban systems*. Englewood Cliffs, N.J.

BERRY, B. J. L. and A. PRED (1965) *Central place studies: a bibliography of theory and applications*. Reg. Sci. Res. Inst. Philadelphia.

BIRD, J. (1963) *The major seaports of the United Kingdom*. London.

BIRD, J. (1968) *Seaport gateways of Australia*. London.

BLAIKIE, P. M. (1971) Spatial organization of agriculture in some north Indian villages. Parts I and II. *Trans. Inst. Brit. Geogr.* 52 and 53, 1–40 and 15–30.

BLUNDEN, W. R. (1971) *The land use transport system: analysis and synthesis*. Oxford.

BORCHERT, J. R. (1961) The twin cities urbanized area: past, present, future. *Geogr. Rev.* 51, 47–70.

BRADNOCK, R. (forthcoming) India: a case study in development planning, in B. W. HODDER and A. M. O'CONNOR (eds.) *Development planning: case studies*. London.

BRECHLING, F. (1967) Trends and cycles in British regional unemployment. *Oxford Econ. Pap.*, New Ser., 19, 1–21.

BRITTON, D. K., B. E. CRACKNELL and I. M. T. STEWART (1969) *Cereals in the United Kingdom: production, marketing and utilization*. Oxford.

BRITTON, J. N. H. (1967) *Regional analysis and economic geography: a case study of manufacturing in the Bristol region*. London.

BROOKFIELD, H. C. (1973) On one geography and a Third World. *Trans. Inst. Brit. Geogr.* 58, 1–20.

BROWN, A. J. (1972) *The framework of regional economics in the United Kingdom.* Nation. Inst. of Econ. and Soc. Res., Econ. and Soc. Stud., 27. Cambridge.

BROWN, R. (1930) *Economic geography.* New York.

BUCHANAN, K. (1970) *The transformation of the Chinese earth.* London.

BUCHANAN, R. O. (1935) *The pastoral industries of New Zealand: a study in economic geography.* Inst. Brit. Geogr. Publ. Number 2.

BUNGE, W. (1966) *Theoretical geography.* Lund Stud. in Geogr., Ser. C, Gen. and Maths. Geogr., 1.

BURGHARDT, A. F. (1972) Income density in the United States. *Ann. Assoc. Am. Geogr.* 62, 455–60.

BURTON, I. (1963) *Accessibility in northern Ontario: an application of graph theory to a regional highway network.* Ontario.

BURTON, I. and R. W. KATES, eds. (1965) *Readings in resource management and conservation.* Chicago.

BUZZACOTT, K. L. (1972) *London's markets: their growth, characteristics and functions.* Unpubl. Ph.D. thesis, Univ. of London.

CAMPBELL, W. J. and M. CHISHOLM (1970) Local variations in retail grocery prices. *Urb. Stud.* 7, 76–81.

CARTER, W. H. and R. E. DODGE (1939) *Economic geography.* New York.

CENTRAL STATISTICAL OFFICE (1968) *Standard industrial classification.* H.M.S.O., London.

CENTRAL STATISTICAL OFFICE (1976) *National income and expenditure.* H.M.S.O., London.

CHISHOLM, G. G. (1889, 1975) *Chisholm's handbook of commercial geography.* 1st and 19th editions. London.

CHISHOLM, M. (1966, 1970) *Geography and economics.* 1st and 2nd editions. London.

CHISHOLM, M. (1968) *Rural settlement and land use.* 2nd edition. London.

CHISHOLM, M. (1971a) Forecasting the generation of freight traffic in Great Britain, in M. CHISHOLM, A. E. FREY and P. HAGGETT (eds.) *Regional Forecasting.* Proc. of the twenty-second symposium of the Colston Res. Soc. London.

CHISHOLM, M. (1971b) Freight transport costs, industrial location and regional development, in M. CHISHOLM and G. MANNERS (eds.) *Spatial policy problems of the British economy.* London.

CHISHOLM, M. and G. MANNERS (1971) Geographical space: a new dimension of public concern and policy, in M. CHISHOLM and G. MANNERS (eds.) *Spatial policy problems of the British economy.* London.

CHISHOLM, M. and J. OEPPEN (1973) *The changing pattern of employment.* London.

CHISHOLM, M. and P. O'SULLIVAN (1973) *Freight flows and the British economy*. London.

CHRISTALLER, W. (1933, 1966) *Die zentralen Orte in Süddeutschland*. Translated by C. W. BASKIN (1966) as *Central places in southern Germany*. Englewood Cliffs, N.J.

CLARK, C. (1960) *The conditions of economic progress*. 3rd edition. London.

CLARK, C. and G. H. PETERS (1965) The intervening opportunities method of traffic analysis. *Traffic Q*. 19, 101–19.

CLARK, W. A. V. (1968) Consumer travel patterns and the concept of range. *Ann. Assoc. Am. Geogr*. 58, 386–96.

CLARK. W. A. V. and G. RUSHTON (1970) Models of intra-urban consumer behaviour and their implications for central place theory. *Econ. Geogr*. 46, 486–97.

COATES, B. E. (1965) The origin and distribution of markets and fairs in medieval Derbyshire. *Derbyshire. Archaeological Journal* 85, 92–111.

COATES, B. E. and E. M. RAWSTRON (1971) *Regional variations in Britain*. London.

CODDINGTON, A. (1970) The economics of ecology. *New Soc*. 22, 9 April, 595–7.

COLE, J. P. (1965) *Latin America: an economic and social geography*. London.

COLE, J. P. and F. C. GERMAN (1972) *A geography of the U.S.S.R.: the background to a planned economy*. 2nd edition. London.

COMMISSION ON THE THIRD LONDON AIRPORT (1971) *Report*. H.M.S.O. London.

CONNELL, J. (1971) The geography of development. *Area* 3, 259–265.

COPPOCK, J. T. (1971) *An agricultural geography of Great Britain*. London.

COURTENAY, P. P. (1972) *A geography of trade and development in Malaya*. London.

DAVIES, B. (1968) *Social needs and resources in local services*. London.

DAVIES, R. L. (1969) Effects of consumer income differences on shopping movement. *Tijdschr. Econ. Soc. Geogr*. 60, 111–21.

DAVIES, R. (1973) The location of service activities, in M. CHISHOLM and B. RODGERS (eds.) *Studies in human geography*. London.

DAWSON, J. A. (1973) Marketing, in J. A. DAWSON and J. C. DOORNKAMP (eds.) *Evaluating the human environment: essays in applied geography*. London.

DAY, R. A. (1973) Consumer shopping behaviour in a planned urban environment. *Tijdschr. Econ. Soc. Geogr*. 64, 77–85.

DENMAN, D. R. and S. PRODANO (1972) *Land-use: an introduction to proprietary land-use analysis*. London.

DEPARTMENT OF EMPLOYMENT AND PRODUCTIVITY (annual) *Family expenditure survey*. H.M.S.O. London.

DICKINSON, R. E. (1934) The markets and market areas of East Anglia. *Econ. Geogr.* 10, 172–82.

DIENES, L. (1972) Investment priorities in Soviet regions. *Ann. Assoc. Am. Geogr.* 62, 437–54.

DONNISON, D. and D. EVERSLEY, eds. (1973) *London–urban patterns, problems and policies*. London.

DUNN, E. S. (1954) *The location of agricultural production*. Gainesville.

DUNNING, J. H. and E. V. MORGAN, eds. (1971) *An economic study of the City of London*. Toronto.

DUTT, A. K. (1969) Intra-city hierarchy of central places: Calcutta as a case study. *Prof. Geogr.* 21, 18–22.

DUTT, D. A. (1965) *Shopping patterns in Calcutta*.

DZIEWONSKI, K. and M. JERCZYNSKI (1971) *Urban-economic base and the functional structure of cities*. Geogr. Stud. No. 87, Inst. Geogr. Polskiej Akademii Nauk, Warsaw.

ELIOT-HURST, M. E. (1970) An approach to the study of non residential land use traffic generation. *Ann. Assoc. Am. Geogr.* 60, 153–73.

ELIOT-HURST, M. E. (1972) *A geography of economic behaviour*. California.

EPSTEIN, T. S. (1962) *Economic development and social change in south India*. Manchester.

ESTALL, R. C. (1966) *New England: a study in industrial adjustment*. London.

ESTALL, R. (1977) *A modern geography of the United States*. 2nd edition. Harmondsworth.

ESTALL, R. C. (1972) Some observations on the internal mobility of investment capital. *Area* 4, 193–8.

ESTALL, R. C. and R. O. BUCHANAN (1973) *Industrial activity and economic geography*. 3rd edition, London.

EVANS, A. W. (1973) *The economics of residential location*. London.

EVERSLEY, D. E. C. (1972a) Old cities, falling populations and rising costs. *G.L.C. Intelligence Unit Q. Bull.* No. 18, 5–17.

EVERSLEY, D. E. C. (1972b) Urban problems in Britain today. *G.L.C. Intelligence Unit Q. Bull.* No. 19, 43–51.

EVERSLEY, D. E. C. (1972c) Rising costs and static incomes: some economic consequences of regional planning in London. *Urb. Stud.* 9, 347–68.

EYLES, J. (1971) Pouring new sentiments into old theories: how else can we look at behavioural patterns? *Area* 3, 242–50.

FARMER, B, H. (1957) *Pioneer peasant colonization in Ceylon: a study in Asian agrarian problems*. London.

FIREY, W. (1960) *Man, mind and land: a theory of resource use.* Chicago.

FIRTH, R. (1952) *Malay fishermen: their peasant economy.* London.

FISHER, A. G. B. (1952) A note on tertiary production, *Econ. J.*, 42, 820–34.

FISHER, C. A. (1948) Economic geography in a changing world. *Trans. Inst. Brit. Geogr.* 14, 69–84.

FLEMING, M. (1969) *Introduction to economic analysis.* London.

FOGEL, R. W. (1964) *Railroads and American economic growth: essays in econometric history.* Johns Hopkins Press, Baltimore.

FOUND, W. C. (1971) *A theoretical approach to rural land-use patterns.* London.

FOX, H. S. A. (1970) Going to market in 13th century England. *Geogr. Mag.*, 42, 658–67.

FRYER, D. W. (1965) *World economic development.* New York.

GALBRAITH, J. K. (1967) *The new industrial state.* London.

GARNER, B. (1967) Models of urban geography and settlement location, in R. J. CHORLEY and P. HAGGETT (eds.) *Models in geography.* London.

GARNER, B. J. (1970) Towards a better understanding of shopping patterns, in R. H. OSBORNE, F. A. BARNES and J. C. DOORNKAMP (eds.) *Geographical essays in honour of K. C. Edwards.* Univ. of Nottingham, Dept. Geogr.

GARRISON, W. L. (1959, 1960) Spatial structure of the economy. *Ann. Assoc. Am. Geogr.* 49, 232–9; 49, 471–82; 50, 357–73.

GARRISON, W. L. (1960) Connectivity of the interstate highway system. *Pap. and Proc. Reg. Sci. Assoc.* 6, 121–37.

GARRISON, W. L. (1966) Solving urban transportation problems. Address to the Annual Conference of Mayors, Dallas, June. Cited by B. J. L. BERRY and F. E. HORTON *Geographic perspectives on urban systems.* Englewood Cliffs, N.J.

GARRISON, W. L., B. J. L. BERRY, D. F. MARBLE, J. D. NYSTUEN and R. L. MORRILL (1959) *Studies of highway development and geographic change.* Seattle.

GARRISON, W. L. and D. F. MARBLE (1961) The structure of transportation networks. *U.S. Department of Commerce, Office of Technical Services*, Washington D.C.

GAUTHIER, H. L. (1968) Transportation and the growth of the Sao Paulo economy. *J. Reg. Sci.* 8, 77–94.

GAUTHIER, H. L. (1970) Geography, transportation and regional development. *Econ. Geogr.* 46, 612–19.

GILBERT, A. (1971) Some thoughts on the 'new geography' and the study of 'development'. *Area 3*, 123–8.

GILBERT, A. and D. E. GOODMAN (1976) Regional income disparities and economic development: a critique in A. Gilbert (ed.) *Development planning and spatial structure*. London.

GINSBURG, N. S. ed. (1960) *Essays on geography and economic development*. Univ. of Chicago, Dept. Geogr. Res. Pap. 62, Chicago

GINSBURG, N. S. ed. (1961) *Atlas of economic development*. Chicago.

GINSBURG, N. (1967) Foundations of economic geography series, in B. J. L. BERRY *Geography of market centres and retail distribution*. Englewood Cliffs, N.J.

GINSBURG, N. (1969) Tasks of geography. *Geogr.* 54, 401–9.

GOLDSMITH, E. *et al.* (1972) *A blueprint for survival*. London.

GOLLEDGE, R. G. (1970) Some equilibrium models of consumer behaviour. *Econ. Geogr.* 46, 417–24.

GOLLEDGE, R. G., G. RUSHTON and W. A. V. CLARK (1966) Some spatial characteristics of Iowa's dispersed farm population and their implications for the grouping of central place functions. *Econ. Geogr.* 42, 261–72.

GOODALL, B. (1972) *The economics of urban areas*. Oxford.

GÖTZ, W. (1882) Noted in many sources, though the original reference seems obscure. See Wooldridge, S. W. and East, W. G. *The spirit and purpose of geography*. London, 1966, 91–2.

GREENHUT, M. L. (1956) *Plant location in theory and practice*. Chapel Hill, N. Carolina.

GREGOR, H. F. (1963) *Environment and economic life*. Princeton, N.J.

GREGOR, H. F. (1970) *Geography of agriculture: themes in research*. Englewood Cliffs, N.J.

GUYOL, N. B. (1971) *Energy in the perspective of geography*. Englewood Cliffs, N.J.

HÄGERSTRAND, T. (1957) *Migration and area: survey of a sample of Swedish migration fields and hypothetical considerations on their genesis*. Lund Studies in Geography, Series B, Human Geography 13, 27–158.

HÄGERSTRAND, T. (1967) *Innovation diffusion as a spatial process*. (Trans. A. Pred) Chicago.

HÄGERSTRAND, T. (1970) What about people in regional science? *Pap. and Proc. Reg. Sci. Assoc.* 24, 7–21.

HAGGETT, P. (1965) *Locational analysis in human geography*. London.

HAGGETT, P. (1967) Network models in geography, in R. J. CHORLEY and P. HAGGETT (eds.) *Models in geography*. London.

HAGGETT, P. (1970) Changing concepts in economic geography, in R. J. CHORLEY and P. HAGGETT (eds.) *Frontiers in geographical teaching*. 2nd edition, London.

HAGGETT, P. (1971) Leads and lags in inter-regional systems: a study of the cyclic fluctuations in the south-west economy, in M. CHISHOLM and G. MANNERS (eds.) *Spatial policy problems of the British economy.* Cambridge.

HAGGETT, P. (1972) *Geography: a modern synthesis.* London.

HAGGETT, P. and R. J. CHORLEY (1969) *Network analysis in geography.* London.

HALL, P. G. (1962) *The industries of London since 1861.* London.

HALL, P. G. ed. (1966) *Isolated state.* An English edition of J. H. von Thünen (1875) *Der isolierte Staat*, translated by C. M. WARTENBURG and P. G. HALL. Oxford.

HALL, P. (1969) Transportation. *Urb. Stud.* 6, 408–35.

HAMILTON, F. E. I. (1968) *Yugoslavia: patterns of economic activity.* London.

HAMILTON, F. E. I. (1970) Planning the location of industry in East Europe: the principles and their impact. *Econ. Plan.* 6, 3–7.

HAMILTON, F. E. I. (1971) Decision-making and industrial location in Eastern Europe. *Trans. Inst. Brit. Geogr.* 52, 77–94.

HARRIS, C. P. and A. P. THIRLWALL (1968) Interregional variations in cyclical sensitivity to unemployment in the U.K. 1949–1964. *Bull. of the Oxford Univ. Inst. of Econ. and Statistics.* 30, 55–66.

HARVEY, D. (1966) Theoretical concepts and the analysis of agricultural land use patterns. *Ann. Assoc. Am. Geogr.* 56, 361–74.

HARVEY, D. (1967) *Behavioural postulates and the construction of theory in human geography.* Univ. of Bristol, Dept. Geogr. Seminar Pap. Ser., Ser. A: No. 6.

HARVEY, D. (1969) *Explanation in geography.* London.

HARVEY, D. (1973) *Social justice and the city.* London.

HAY, A. M. (1971) Notes on the economic basis for periodic marketing in developing countries. *Geographical Analysis*, 3, 4, 393–401.

HAY, A. M. (1973) *Transport for the space economy.* London.

HAY, A. M. and R. H. T. SMITH (1970) *Interregional trade and money flows in Nigeria, 1964.* Ibadan.

HECOCK, R. D. and J. F. ROONEY (1968) Towards a geography of consumption. *Prof. Geogr.* 20, 392–5.

HILL, P. (1963) Markets in Africa. *J. Mod. Afr. Stud.* 1, 441–53.

HILL, P. (1972) *Rural Hausa: a village and a setting.* Cambridge.

HILLING, D. (1969) The evolution of the major ports of West Africa. *Geogr. J.* 135, 365–77.

HIRSCHMANN, A. O. (1958) *The strategy of economic development.* London.

HOCH, I. (1969) *Progress in urban economics.* Resources for the Future, Washington D.C.

HODDER, B. W. (1964) Some comments on the origins of traditional markets in Africa South of the Sahara. *Trans. Inst. Brit. Geogr.* 36, 95–105.

HODDER, B. W. (1973) *Economic development in the tropics*. 2nd edition, London.

HODDER, B. W. and U. I. UKWU (1969) *Markets in West Africa*. Ibadan.

HOLLY, B. P. and J. O. WHEELER (1972) Patterns of retail location and the shopping trips of low income households. *Urb. Stud.* 9, 215–20.

HOPE, R. (1969) *Economic geography*. 5th edition, London.

HOTELLING, H. (1929) Stability in competition. *Econ. J.* 39, 41–57.

HOUSE, J. W. (1969) *Industrial Britain: the North East*. Newton Abbot.

HOUSE, J. W. ed. (1974) *The U.K. space*. London.

HOYLE, B. S. (1968) East African seaports: an application of the concept of 'Anyport'. *Trans. Inst. Brit. Geogr.* 44,163–83.

HOYLE, B. S. ed. (1973) *Transport and development*. London.

HUDSON, R. (1971) *Towards a theory of consumer spatial behaviour*. Univ. of Bristol, Dept. Geogr. Seminar Pap. Ser., Ser. A: No. 22.

HUFF, D. L. (1960) A topographical model of consumer space preferences. *Pap. and Proc. Reg. Sci. Assoc.* 6, 159–74.

HUMPHRYS, G. (1972) with a contribution by J. W. ENGLAND. *Industrial Britain: South Wales*. Newton Abbot.

ISARD, W. (1956) *Location and space economy: a general theory relating to industrial location, market areas, land use, trade and urban structure*. Cambridge.

ISARD, W. (1968) Some notes in the linkage of the ecologic and economic systems. *Pap. and Proc. Reg. Sci. Assoc.* 22, 85–96.

ISARD, W., D. F. BRAMHALL, G. A. P. CARROTHERS, J. H. CUMBERLAND, L. N. MOSES, D. O. PRICE, and E. W. SCHOONER (1960) *Methods of regional analysis: an introduction to regional science*. New York.

ISARD, W. et al. (1972) *Ecologic-economic analysis for regional development: some initial explorations with particular reference to recreational resource use and environmental planning*. New York.

ISARD, W. and R. E. KUENNE (1953) The impact of steel upon the greater New York–Philadelphia industrial region. *Rev. Econ. and Stat.* 21, 289–301.

JACKSON, R. (1971) Periodic markets in southern Ethiopia. *Trans. Inst. Brit. Geogr.*, 53, 31–42.

JANELLE, D. G. (1969) Spatial reorganization: a model and a concept. *Ann. Assoc. Am. Geogr.* 59, 348–64.

JEANS, D. N. (1967) Competition, momentum and inertia in the location of commercial institutions: case studies in some London commodity markets. *Tijdschr. Econ. Soc. Geogr.* 58, 11–19.

JENSEN, R. G. (1969) Regionalization and price zonation in Soviet agricultural planning. *Ann. Assoc. Am. Geogr.* 59, 324–47.

JOHNSTON, R. J. (1973) *Spatial structures*. London.

JOHNSTON, W. B. ed. (1965) *Traffic in a New Zealand city*. Christchurch.

JONES, G. E. (1967) The adoption and diffusion of agricultural practices. *Wld. Agric. Econ. and Rur. Sociol. Abs.* 9, 1–34.

KANSKY, K. J. (1963) *Structure of transport networks: relationships between network geometry and regional characteristics*. Univ. of Chicago, Dept Geogr. Res. Pap. 84.

KASPERSON, R. E. and J. V. MINGHI eds. (1970) *The structure of political geography*. London.

KEEBLE, D. E. (1967) Models of economic development, in R. J. CHORLEY and P. HAGGETT (eds.) *Models in geography*. London.

KELLETT, J. R. (1969) *The impact of railways on Victorian cities*. London.

KINDLEBERGER, C. P. (1963) *International economics*. Homewood.

KING, L., E. CASSETTI and D. JEFFREY (1969) Economic impulses in a regional system of cities: a study of spatial interaction. *Reg. Stud.* 3, 213–18.

KING, L. J., E. CASSETTI and D. JEFFREY (1972) Cyclical fluctuations in unemployment levels in U.S. metropolitan areas. *Tijdschr. Econ. Soc. Geogr.* 53, 345–52.

KIRK, W. (1951) Historical geography and the concept of the behavioural environment. *Ind. Geogr. J., Silver Jubilee Volume*, 152–160.

KIRK, W. (1963) Problems of geography. *Geogr.* 48, 357–71.

KOLARS, J. and R. MALIN (1970) Population and accessibility: an analysis of Turkish railroads. *Geogr. Rev.* 60, 229–46.

KRUMME, G. (1969) Toward a geography of enterprise. *Econ. Geogr.* 45, 30–40.

LACHENE, R. (1965) Networks and the location of economic activities. *Pap. and Proc. Reg. Sci. Assoc.* 14, 183–96.

LAKSHMANAN, T. R. and W. C. HANSEN (1965) A retail market potential model. *J. of the Am. Inst. of Planners*, 31, 134–43.

LAMARTINE-YATES, P. (1959) *Forty years of foreign trade*. London.

LEONTIEFF, W. W. (1954) Domestic production and foreign trade: the American capital position re-examined. *Economia Internazionale*, 7, 3–32.

LEWIS, P. W. and P. N. JONES (1970) *Industrial Britain: the Humberside region*. Newton Abbot.

LEWIS, W. A. (1959) On assessing a development plan. *Econ. Bull. Ghana*, 3, 6–7.

LINDER, S. B. (1961) *An essay on trade and transformation*. New York.

LINGE, G. J. R. (1971) Government and spatial behaviour, in G. J. R. LINGE and P. J. RIMMER (eds.) *Government influence and the location of economic activity*. Canberra.

LINNEMANN, H. (1966) *An econometric study of international trade flows*. Amsterdam.

LIPSEY, R. G. (1975) *An introduction to positive economics*. 4th edition. London.

LLOYD, P. E. and P. DICKEN (1972) *Location in space: a theoretical approach to economic geography*. London.

LÖSCH, A. (1967) *The economics of location*. Translated by W. H. WOGLOM and W. F. STOLPER. New York.

MCCARTY, H. H. and J. B. LINDBERG (1966) *A preface to economic geography*. Englewood Cliffs, N.J.

MCCONNELL, J. E. (1970) A note on the geography of commodity trade. *Prof. Geogr.* 22, 181–4.

MCDANIEL, R. and M. E. ELIOT-HURST (1968) *A systems analytic approach to economic geography*. Publ. No. 8, Commis. on Coll. Geogr., Assoc. Am. Geogr., Washington D.C.

MCNEE, R. B. (1958) Functional geography of the firm, with an illustrative case study from the petroleum industry. *Econ. Geogr.* 34, 321–337.

MCNEE, R. B. (1960) Toward a more humanistic economic geography: the geography of enterprise. *Tijdschr. Econ. Soc. Geogr.* 51, 201–6.

MANNERS, G. (1969) New resource evaluations, in R. U. COOKE and J. H. JOHNSON (eds.) *Trends in geography*. Oxford.

MANNERS, G. (1970) Greater London Development Plan: location policy for manufacturing industry. *Area* 3, 54–6.

MANNERS, G. (1971a) *The changing world market for iron ore 1950–1980: an economic geography*. Resources for the Future. London.

MANNERS, G. (1971b) *The geography of energy*. 2nd edition, London.

MANNERS, G. ed. (1972) *Regional development in Britain*. London.

MARBLE, D. F. (1966) A theoretical exploration of individual travel behaviour, in W. L. GARRISON and D. F. MARBLE (eds.) *Quantitative geography, Part I: economic and cultural topics*. Northwestern Univ. Stud. in Geogr., Evanston.

MARTIN, J. E. (1966) *Greater London, an industrial geography*. London.

MEAD, W. R. (1959) *An economic geography of the Scandinavian states and Finland*. London.

MEADOWS, D. H. et al. (1972) *The limits to growth*. London.

MEINIG, D. W. (1962) A comparative historical geography of two railnets: Columbia basin and South Australia. *Ann. Assoc. Am. Geogr.* 52, 394–413.

MELAMID, A. (1962) Geography of the world petroleum price structure. *Econ. Geogr.* 38, 283–98.

MILIBAND, R. (1969) *The state in capitalist society*. London.

MINISTRY OF TRANSPORT (1963) *Traffic in towns: a study of the long term problems of traffic in urban areas*. H.M.S.O., London.

MITCHELL, R. and C. RAPKIN (1954) *Urban traffic: a function of land use.* New York.

MORGAN, W. B. and R. J. C. MUNTON (1971) *Agricultural geography.* London.

MORRILL, R. L. and E. H. WOHLENBERG (1971) *The geography of poverty in the United States.* New York.

MOUNTJOY, A. B. (1966) *Industrialization and underdeveloped countries.* 2nd edition, London.

MOUNTJOY, A. B. ed. (1971) *Developing the underdeveloped countries.* London.

MUNTON, R. J. C. (1969) The economic geography of agriculture, in R. U. COOKE and J. H. JOHNSON (eds.) *Trends in geography.* Oxford.

MURDIE, R. A. (1965) Cultural differences in consumer travel. *Econ. Geogr.* 41, 211–33.

MYRDAL, G. (1957) *Economic theory and underdeveloped regions.* London.

NADER, G. A. (1969) Socio-economic status and consumer behaviour. *Urb. Stud.* 6, 235–45.

NEWBIGIN, M. I. (1928) *Commercial geography.* London.

O'CONNOR, A. M. (1965a) New railway construction and the pattern of economic development in East Africa. *Trans. Inst. Brit. Geogr.* 36, 21–30.

O'CONNOR, A. M. (1965b) *Railways and development in Uganda: a study in economic geography.* East Afr. Inst. of Soc. Res., East Afr. Stud. 18, Nairobi.

O'CONNOR, A. M. (1968) *An economic geography of East Africa.* London.

ODELL, P. R. (1963) *An economic geography of oil.* London.

ODELL, P. R. (1969) *Natural gas in Western Europe: a case study in the economic geography of energy resources.* Inaugural address, Netherlands Schl of Econ., Rotterdam, May 29th. Haarlem.

ODELL, P. R. (1975) *Oil and world power: a geographical interpretation.* 4th edition. Harmondsworth.

O'FARRELL, P. N. and M. A. POOLE (1972) Retail grocery price variation in Northern Ireland. *Reg. Stud.* 6, 83–92.

OHLIN, B. G. (1935) *Interregional and international trade.* Harvard Univ. Press.

ORD, H. W. and I. LIVINGSTONE (1969) *An introduction to West African economies.* London.

O'RIORDAN, T. (1971) *Perspectives on resource management.* London.

O'SULLIVAN, P. M. (1969) *Transport networks and the Irish economy.* London Schl. of Econ. and Pol. Sci., Geogr. Paper 3.

O'SULLIVAN, P. (1970) Variations in distance friction in Great Britain. *Area* 2, 36–9.

OWENS, D. (1968) *Estimates of the proportion of space occupied by roads and footpaths in towns.* Road Res. Labor., Report LR154.

PAHL, R. E. (1968) *Spatial structure and social structure.* Centre for Env. Stud., Working Pap. 10.

PAHL, R. (1971) Poverty and the urban system, in M. CHISHOLM and G. MANNERS (eds.) *Spatial policy problems of the British economy.* London.

PARKER, G. (1968) *The logic of unity: an economic geography of the Common Market.* London.

PARSONS, G. F. (1972) The giant manufacturing corporations and balanced regional growth in Britain. *Area* 4, 99–103.

PATERSON, J. H. (1972) *Land, work and resources: an introduction to economic geography.* London.

PATMORE, J. A. (1966) The contraction of the network of railway passenger services in England and Wales 1836–1962. *Trans. Inst. Brit. Geogr.* 38, 105–18.

PENROSE, E. T. (1968) with a chapter by P. R. ODELL *The large international firm in developing countries: the international petroleum industry.* London.

PERLE, E. D. (1964) *The demand for transportation: regional and commodity studies in the United States.* Univ. of Chicago, Dept. Geogr. Res. Pap. 95.

PERLOFF, H. S. and L. WINGO eds. (1968) *Issues in urban economics.* Resources for the Future, Baltimore.

PERROUX, F. (1955) Note sur la notion de 'pole de croissance'. *Economie Appliqué* 8, 307–20.

PIRENNE, H. (1965) *Economic and social history of medieval Europe.* Translated from the French by I. E. CLEGG. 8th edition, London.

POLANYI, K., C. M. ARENSBURG and H. W. PEARSON eds. (1957) *Trade and markets in the early empires: economies in history and theory.* London.

POOLE, M. A. (1968) *An analysis of rural electricity consumption patterns in the Republic of Ireland.* Unpubl. Ph.D. thesis, Queen's Univ., Belfast.

POOLE, M. A. (1970) Rural domestic electricity consumption in the Republic of Ireland: an exploratory case study in consumption geography. *Ir. Geogr.* 6, 113–35.

PREBISCH, R. (1950) *The economic development of Latin America and its principal problems.* United Nations, New York.

PRED, A. (1967) *Behaviour and location: foundations for a geographic and dynamic location theory.* Parts I and II. Lund Stud. in Geogr., Ser. B, Human Geogr. No. 27.

PRESCOTT, J. R. V. (1968) *The geography of state policies.* London.

PRESCOTT, J. R. V. (1972) *Political geography.* London.

RAWSTRON, E. M. (1958) Three principles of industrial location. *Trans. Inst. Brit. Geogr.* 25, 135–42.

RAY, D. M. (1967) Cultural differences in consumer travel behaviour in Eastern Ontario. *Can. Geogr.* 11, 143–56.

REILLY, W. J. (1929) Methods for the study of retail relationships. *Univ. of Texas Bull.* 2944.

RICHARDSON, H. W. (1969) *Regional economics: location theory, urban structure and regional change.* London.

RICHARDSON, H. W. (1971) *Urban economics.* Harmondsworth.

RICHARDSON, H. W. (1972) *Input-output and regional economics.* London.

ROBSON, B. T. (1969) *Urban analysis: a study of city structure with special reference to Sunderland.* London.

RODGERS, A. L. (1952) Industrial inertia: a major factor in the location of the steel industry in the U.S. *Geogr. Rev.* 42, 56–66.

ROSTOW, W. W. (1960) *The stages of economic growth: a non communist manifesto.* Cambridge.

ROWLEY, G. (1972) Spatial variations in the prices of central goods: a preliminary investigation. *Tijdschr. Econ. Soc. Geogr.* 63, 360–8.

RUSHTON, G. (1969) Analysis of spatial behaviour by revealed space preference. *Ann. Assoc. Am. Geogr.* 59, 391–400.

RUSHTON, G., R. G. GOLLEDGE and W. A. V. CLARK (1967) Formulation and test of a normative model for the spatial allocation of grocery expenditures by a dispersed population. *Ann. Assoc. Am. Geogr.* 57, 389–400.

RUTHERFORD, J., M. I. LOGAN and G. J. MISSEN (1966) *New viewpoints in economic geography.* Sydney.

SAMUELSON, P. A. (1977) *Economics: an introductory analysis.* 10th edition. New York.

SANT, M. (1973) *The geography of business cycles.* London Schl. of Econ. and Pol. Sci., Geogr. Paper 5.

SCHILLER, R. K. (1972) The measurement of the attractiveness of shopping centres to middle class luxury consumers. *Reg. Stud.* 6, 291–297.

SCOTT, A. J. (1971) *An introduction to spatial allocation analysis.* Commis. on Coll. geogr. Resource Pap. No. 9, Assoc. Am. Geogr., Washington D.C.

SIMMONS, I. G. (1966) Ecology and land use. *Trans. Inst. Brit. Geogr.* 38, 59–72.

SIMMONS, I. G. (1973) Conservation, in J. A. DAWSON and J. C. DOORNKAMP (eds.) *Evaluating the human environment: essays in applied geography.* London.

SIMON, H. A. (1952) A behavioural model of rational choice. *Q. J. of Econ.* 69, 99–118.

SIMPSON, E. S. (1966) *Coal and the power industries in post-war Britain.* London.

SKINNER, G. W. (1964–5) Marketing and social structure in rural China. *J. of Asian Stud.* 34, 3–43; 198–228; 363–99.

SLATER, D. (forthcoming) Peru, in B. W. HODDER and A. M. O'CONNOR (eds.) *Development Planning: case studies.* London.

SMAILES, A. E. (1964) Foreword, in D. GROVE and L. HESZAR *The towns of Ghana: the role of service centres in regional planning.* Accra.

SMITH, D. A. (undated) quoted by E. J. TAAFE and H. L. GAUTHIER (1973) *Geography of transportation.* Englewood Cliffs, N.J. 80–2.

SMITH, D. M. (1966) A theoretical framework for geographical studies of industrial location. *Econ. Geogr.* 42, 95–113.

SMITH, D. M. (1969) *Industrial Britain: the North West.* Newton Abbot.

SMITH, D. M. (1971) *Industrial location: an economic geographical analysis.* New York.

SMITH, D. M. (1972) Towards a geography of social well-being: interstate variations in the United States, in R. PEET (ed.) *Geographical perspectives on American poverty.* Antipode monog. in soc. geogr., No. 1, Worcester.

SMITH, D. M. (1973a) *The geography of social well-being in the United States.* New York.

SMITH, D. M. (1973b) *An introduction to welfare geography.* Univ. of the Witwatersrand, Dept. of Geogr. Occ. Paper 3.

SMITH, R. H. T. (1964a) Development and function of transport routes in southern New South Wales 1860–1930. *Aust. Geogr. Stud.* 2, 47–65.

SMITH, R. H. T. (1964b) Toward a measure of complementarity. *Econ. Geogr.* 40, 1–8.

SMITH, R. H. T. (1970) Concepts and methods in commodity flow analysis. *Econ. Geogr.* 46 (supplement), 404–16.

SMITH, R. H. T. (1971) Market periodicity and locational patterns in West Africa, in C. MEILASSOUX (ed.) *The development of indigenous trade and markets in West Africa.* London.

SMITH, R. H. T., E. J. TAAFFE and L. J. KING eds. (1968) *Readings in economic geography: the location of economic activity.* Chicago.

SMITH, W. (1949) *An economic geography of Great Britain.* London.

SORRE, M. (1962) The geography of diet, in P. L. WAGNER and M. W. MIKESELL (eds.) *Readings in cultural geography.* Univ. of Chicago Press, Chicago.

STARKIE, D. N. M. (1967) *Traffic and industry. A study of traffic generation and spatial interaction.* London Schl. of Econ. and Pol. Sci., Geogr. Pap., 3.

STARKIE, D. N. M. (1973) Transportation planning and public policy. *Progress in Planning,* 1, 4.

STEED, G. P. F. (1971a) Changing processes of corporate environment relations. *Area* 3, 207–11.

STEED, G. P. F. (1971b) Locational implications of corporate organization of industry. *Can. Geogr.* 15, 54–7.

STEUER, M. D., P. ABELL, J. GENNARD, M. PEARLMAN, R. REES, B. SCOTT and K. WALLIS (1973) *The impact of foreign direct investment on the United Kingdom*. Dept. of Trade and Industr., H.M.S.O. London.

STINE, J. H. (1962) Temporary aspects of tertiary production elements in Korea, in F. R. PITTS (ed.) *Urban systems and economic development*. Univ. of Oregon, Schl of Busin. Administration, Eugene.

STOUFFER, S. A. (1940) Intervening opportunities: a theory relating mobility and distance. *Am. Sociol. Rev.* 5, 845–67.

STREETEN, P. (1971) The frontiers of development studies, in I. LIVINGSTONE (ed.) *Economic policy for development*. Harmondsworth.

SWANN, D. (1975) *The economics of the Common Market*. 3rd edition. Harmondsworth.

TAAFFE, E. J. and H. L. GAUTHIER (1973) *Geography of transportation*. Englewood Cliffs, N.J.

TAAFFE, E. J. and L. J. KING (1966) *Networks of cities*. Guidelines, Unit 3 in Limited Schools Trials, High Schl. Geog. Project, Assoc. Am. Geogr.

TAAFFE, E. J., R. L. MORRILL and P. R. GOULD (1963) Transport expansion in underdeveloped countries: a comparative analysis. *Geogr. Rev.* 27, 240–54.

TAAFFE, R. N. (1960) *Rail transportation and the economic development of Soviet Central Asia*. Univ. of Chicago, Dept. Geogr. Res. Pap. 64.

TAAFFE, R. N. (1962) Transportation and regional specialization: the example of Soviet Central Asia. *Ann. Assoc. Am. Geogr.* 52, 80–98.

TANNER, W. F. and A. F. WILLIAMS (1967) Port development and planning strategy. *J. of Transp. Econ. and Policy*. 1, 315–24.

TAYLOR, M. A. (1968) *Studies of travel in Gloucester, Northampton and Reading*. Road Res. Labor. Rep. LR141.

TEGSJO, B. and S. OBERG (1966) Concept of potential applied to price formation. *Geografiska Ann.* 48B, 51–8.

THOMAN, R. S. and E. C. CONKLIN (1967) *Geography of international trade*. Englewood Cliffs, N.J.

THOMAN, R. S., E. C. CONKLIN and M. H. YEATES (1968) *The geography of economic activity*. 2nd edition, New York.

THOMPSON, I. B. (1970) *Modern France: a social and economic geography*. London.

THOMPSON, W. R. (1965) *A preface to urban economics*. Resources for the future. Baltimore.

THORPE, D. and G. A. NADER (1967) Customer movement and shopping centre structure: a study of a central place system in northern Durham. *Reg. Stud.* 1, 173–91.

TINBERGEN, J. (1962). *Shaping the world economy*. New York.

TÖRNQVIST, G. (1970) *Contact systems and regional development*. Lund Stud. in Geogr., Ser. B, 35.

TOWNROE, P. (1971) *Industrial location decisions. A study in management behaviour*. Univ. of Birmingham, Centre for Urb. and Reg. Stud., Occas. Pap. 15.

TREGEAR, T. R. (1970) *An economic geography of China*. London.

UKWU, U. I. (1969) The markets of Iboland, in B. W. HODDER and U. I. UKWU *Markets of West Africa*. Ibadan.

ULLMAN, E. L. (1956) The role of transportation and the bases for spatial interaction, in W. L. THOMAS Jr. (ed.) *Man's role in changing the face of the earth*. Univ. of Chicago Press, Chicago.

ULLMAN, E. L. (1957) *American commodity flow*. Univ. of Washington Press, Seattle.

UNITED NATIONS (1976) *Statistical Yearbook*. New York.

UNITED NATIONS (1973) *Yearbook of national accounts statistics*. New York.

URRY, J. and J. WAKEFORD eds. (1973) *Power in Britain*. London.

VANCE, J. E. Jr. (1970) *The merchant's world: the geography of wholesaling*. Englewood Cliffs, N.J.

WARNTZ, W. (1959) *Towards a geography of price: a study in geoeconometrics*. Philadelphia.

WARNTZ, W. (1961) Transatlantic flights and pressure patterns. *Geogr. Rev.* 51, 187–212.

WARNTZ, W. (1965) *Macrogeography and income fronts*. Reg. Sci. Res. Inst., Monog. Ser. 3.

WATTS, H. D. (1971) The location of the beet sugar industry in England and Wales. *Trans. Inst. Brit. Geogr.* 53, 95–116.

WATTS, H. D. (1972) Further observations on regional growth and large corporations. *Area* 4, 269–73.

WEBB, J. W. (1959) Basic concepts in the analysis of small urban centres of Minnesota. *Ann. Assoc. Am. Geogr.* 49, 55–72.

WEBER, A. (1909) *Über den standort der industrien*. Translated by C. J. FRIEDRICH (1929) as *Alfred Weber's theory of the location of industries*. Chicago.

WELLS, S. J. (1969) *International economics*. London.

WERNER, C. (1968) The law of refraction in transportation geography: its multivariate extension. *Can. Geogr.* 12, 28–40.

WHEELER, J. O. (1972) Trip purposes and urban activity linkages. *Ann. Assoc. Am. Geogr.* 62, 641–54.

WHITE, H. P. and M. B. GLEAVE (1971) *An economic geography of West Africa*. London.

WILLIAMSON, J. G. (1965) Regional inequality and the process of national development: a description of the patterns. *Econ. Dev. and Cult. Change.* 13, 3–45.

WILSON, G. W. (1966) Toward a theory of transport and development, in G. W. WILSON, B. R. BERGMAN, L. V. HIRSCH and M. S. KLEIN *The impact of highway investment on development*. The Brookings Institution, Washington D.C.

WILSON, G. W., B. R. BERGMAN, L. V. HIRSCH and M. S. KLEIN (1966) *The impact of highway investment on development*. The Brookings Institution, Washington D.C.

WISE, M. J. (1956) Economic geography and the location problem. *Geogr. J.* 122, 98–100.

WOLPERT, J. (1964) The decision process in spatial context. *Ann. Assoc. Am. Geogr.* 54, 537–58.

WOOD, P. A. (1969) Geography and economic planning, in R. U. COOKE and J. H. JOHNSON (eds.) *Trends in geography*. Oxford.

WOOLDRIDGE, S. W. and W. G. EAST (1966) *The spirit and purpose of geography*. London.

YEATES, M. H. (1968) *An introduction to quantitative analysis in economic geography*. New York.

ZETTEL, R. M. and R. R. CARLL (1962) *Summary review of major metropolitan area transportation studies in the United States*. Inst. of Trans. and Traffic Engineering, Univ. of California.

ZIMMERMAN, E. W. (1933, 1972) *World resources and industry*. Edited by W. N. PEACH and J. A. CONSTANTIN. New York.

Author Index

General Index